The Reset

Melina Bricker

This book is for Ron and Brenda Bricker. You made a man that makes the world better, safer, stronger, and more honest. You placed your family first, a rare occurrence in a world that leans toward selfishness. Your combined strength has grafted onto each of your children, and I have the deep honor of loving and being loved by all of you.

Disclaimer

This book is a work of nonfiction based on research and personal experiences. Some details have been altered to protect the identities of individuals. Any resemblance to specific persons is unintentional.

WAGON RUTS: AN INTRODUCTION

Imagine for a moment that you are born a wagon. Any wagon. Perhaps an Oregon-trail era wagon, completely wooden save some blacksmith elements. Wagons, like people, will have some variation in appearance and operational capability--some people are covered wagons, some people are chariots, some people are fancy Victorian wagons with cushioned seats and glossy exteriors. It really doesn't matter too much about the appearance of your wagon--it just needs to function. Have your wagon in your mind? Good.

Now imagine life as a road stretching in front of you. The first part of the road is early childhood. It's a mucky, wet dirt path that invites slips and falls, requires maneuvering around potholes, and is generally messy as you learn to drive your wagon. You're also not quite sure how to handle it yet--it's your development, your wagon, your genetics, your *self*--but you don't have anything but a general sense that the road stretches out for an unknown amount of time, and you should probably keep moving forward.

You are a wagon, and life is a road. We all share these things. What we do not share, however, is our specific path. This is unique to us, full of turns, and rife with heartache and celebration. What we encounter on our path informs how we drive our wagons. Ideally we have someone with us at first to teach us how to respond to things that pop up. A bear, snarling and puffing, could stand in our path, reared on hind legs, pawing aggressively. What would you do? You'd go around, or speed through it, or freeze to consider options, or perhaps even attempt to throw out provisions and distract it from its ire. Now imagine this bear is so aggressive and damaging, or so constantly popping up, that you keep readjusting your wagon at the slightest hint of it threatening you. In fact, you've gotten so good at managing that bear, you can turn right-quick to the left and circumvent it before it can reach out its paw close enough to hurt you. Or perhaps you have cleverly thought of its favorite treats after years of encounters, so you carry with you an array of offerings just to make it pause its attack as it snaps up your supplies. Perhaps you're more aggressive--you've learned that if you speed up, yell, scream, and attack *first,* the bear leaves you alone.

Stay with me on this. It's a long analogy, but it's worth it. I promise.

Remember, you're a wagon, life's a road, your path's your own, and childhood is mucky and slippery. You've also got a bear on your path, which is a serious pain, but you've found out how to deal with it. As you reach adolescence, you realize driving is way easier. The muck of early childhood is drying out, and by the time you reach adulthood, it's almost concrete in how set it's become.

The only problem now is, you can't really tell what's dangerous, because your whole journey has been trying to avoid, fight, or please *your* bear. Now everyone's a bear, even if they look like a person. So every time they get close, you avoid them, or attack them, or become paralyzed with indecision, or you try to please them, even if it starves you or hurts you. And even if you try to take a different path, you can't--the ruts are dried, and the path is set. Your wheels are no longer in your control.

This is chronic, treatment resistant disorder pathology--and it starts with trauma. Your earliest development is neurobiological preparation and training to navigate the world. This training requires guidance, and grownups are all bears. Your bear should sit with you, guide you, instruct you, protect you, and show you in myriad ways that you are valuable enough to protect, teach, and care about. If your bear doesn't even try to, and just lets you get chewed up and swatted at, or worse, is the one blocking your healthy development, your identity is formed in a state of toxic stress. You are trained to survive, or you are trained to thrive.

The path--your development--is an orchestral interplay of genetics, environment, circumstance, events, rearing, pain, pleasure, nutrition, *life*. Your brain is tasked with interpreting, embodying, and modifying your existence as you develop, and those tasks become increasingly more difficult as you age, with critical plasticity periods in early childhood and adolescence small windows of opportunity to figure the best path forward. When they close, the neural circuitry has limited change, and your ruts become concretized no matter how much you hate their influence on your wagon wheels. This means that even if you love her, you'll still push her away because your ruts say you don't deserve love. It means that you will keep binge eating and purging, because your ruts tell you to consume dopamine-dumping foods only to have shame overwhelm you. You exercise control you never had by throwing up. Or perhaps by restricting eating to a point of pleasure in starvation and pain in even the smallest amount of food. It's so powerful that you still believe there is fat to trim when you see a skin-skeleton staring back at you from the mirror.

This is why trauma creates lifelong disease ranging from diabetes to heart disease, addiction to academic failure, risky sexual behaviors to anorexia, and all the points in between and shared amongst the pages of clinical diagnostic manuals. It has the benefit of providing the pharmaceutical industry endless experimentation and profit, and over-zealous clinicians over-therapizing opportunities that see incremental success while being comforted with recurrent symptoms in their forever-patients.

Profiting from trauma-induced, chronic and treatment-resistant disease is not only sort of icky, it's fundamentally immoral. It's also the primary operation of American health.

What if I told you there was a crazy, unique tool that you could use to address those ruts, those seemingly permanent neurobiological responses? What if I had a magic tool, and someone to help you understand how you use it, and that you could dig into those ruts to prevent your wagon from the same chronic behaviors that keep ruining your path forward?

Psychedelics have been used as long as humans have been around. From cave paintings of magic mushrooms to modern Indigenous ceremonies, psychedelics have been used as tools of community, faith, imagination, and healing. They are the magic tool. With effort and guidance, this tool firmly in your hands allows you to get out of your wagon and really take a look at why the ruts are the way they are. You stop time. You see the bear, walk around it, and realize up close it's this strange, weird likeness to your dad, or your mom, and that really it was a person that should have kept you safe, but instead, opted to hurt you, or abandon you, or insist you are a worthless burden. You see it as it is. And now you have the tool to repair your path.

This is a book about taking back American health through psychedelics. Taking it back from your state-sanctioned, socially-conflated traumas, taking it back from corporations and their policy makers pretending to act in the best interests of our country as they make us fatter, angrier, sadder, and poorer; and living a life of hope and renewal.

It's time for a reset.

1

REMOVE THE PARENTS

American society has always been a strange, tricky little creature. The combined origination principles of independence and efficacious productivity have, over time, augmented the identity of Americans from freedom seekers to machines of industry. While Americans idolize innovation, we have also effectively pushed innovation further from individual reach with dependence upon employer-sponsored healthcare, abysmal wage increases, inflation rates, and enormous income inequality. Innovation is engineered creativity, and requires flexibility in time, critical thought, widely varied approaches to education and learning, and opportunities to pursue new ideas and concepts--all things the majority of Americans cannot access as they're struggling to stretch one paycheck to the next. For many families, when both parents finally drag themselves home from work that they may or may not enjoy, the only bandwidth remaining is used heating up marginally healthy food, hoping the kids are telling the truth about homework completion, and then melting into a soft space to zone out watching a screen. Often, sadly, we are holding a smaller screen as we do so, commenting or reacting to social media posts that deceivingly appear to be the full picture of other people's lives.

Atrophy of meaning is a real issue. A significant one. We are creatures of curiosity, and when we had the marvelous gift of clear skies to gaze upward at, we explored the divine in addition to the earthly. Oral traditions were memory-builders. The process of memory development, storage, and recall is a fascinating area of study--long, ancient histories are woven into the human psyche in a way that is almost unexplainable. Or at least unexplainable without imagination. I think, often, that we hold memories like butterflies do. Butterflies, when presented with a threat, will respond to the same threat as a caterpillar regardless of its ability to hurt them. They remember ideas and instances of significance, even though it cocooned, dissolved into liquid, reformed into a butterfly, and transformed into a completely different life form. Memories are emotional experiences placed into stories, connecting feelings together to create a history of life from the individual to the societal and beyond. I believe firmly that empathy derives from emotional memories of the self, and compassion is derived from emotional memories about another's experience.

But what happens when we stop looking up at the stars? Or when the lights of industry burn so brightly they pollute the imaginative space between us and the heavens? What happens when we stop creating art because we can just download it, or stop creatively writing because AI can tell us a story, or stop being inventive all together, because we're fed entertaining flashes and sounds that sedate our creative pursuits? What happens when we lose the drive of ambition and curiosity, all because we can be temporarily distracted with a click, swipe, snack, drink, or purchase? What happens when our temporary distractions become permanent?

We lose meaning. When we stop looking up, we stop looking in. And *that* is where American society has landed in the 21st Century. We have ceased to look to the heavens for inspiration, to our internal identities for drive and innovation, and we look laterally around us for solutions to problems we simply can't define. Like broken-toothed, starving, geriatric animals, we start trying to eat anything that we can catch hold of, whether it is dangerous, harmful, or poisonous. I read once that poison is anything in too great an amount--compassion is poison when it becomes enabling and codependency. Aggression is poison when it's unwarranted violence. Relaxation is poison when its laziness, and peacefulness is poison when it becomes indifference. The key to having meaning is rooted in purpose from within and a curiosity about the unknown--it requires hope that the universe is full of generative love and connectedness, and belief that you are an accepted, critical part of it all. The well of meaning is poisoned with distraction, consumerism, and forced productivity to contribute to a society that sees workers, not creators. And we keep drinking it, because it looks like the only water source around.

Humans need purpose. We need to know that we have a reason for waking, working, loving, celebrating, eating, all of it. Scarcity is the progenitor of creativity--when you have less, you find ways of making things work to enhance or sustain life. When you have *enough,* you find ways to improve on systems and processes. You can imagine greater things, and you have the time and space to explore your human story and that of those around you through art, invention, community, and faith. You have your needs met--not just the need for safety, nourishment, social interaction, and learning, but the need for purpose as well.

When you have *excess,* you begin to encounter issues. The pursuit of "life, liberty, and property" was the philosophy of Enlightenment-era John Locke, hugely influential to our Founding Fathers. It was later changed in the American zeitgeist by Jefferson to "life, liberty, and the pursuit of happiness." An interesting exchange of words that now, I fear, is changed further to "life, liberty, and the pursuit of excess." We lost, somewhere, our sense of meaning and purpose, and filled it instead with things we want to consume--media, entertainment, processed foods, wealth without limit, fame and infamy...all things, you'll note, that come from external sources on the lateral, Earthly plane. All impermanent, devoid of nourishment, moved to excess as a result of addiction to trying to feel better. It's poison.

The women's movements establishing voting rights, independent bank account ownership, wearing pants instead of skirts, equal employment, and bodily autonomy provide women in the United States with the ability to embody that founding principle: life, liberty, personal pursuits. One issue is, however, that this well is open to poison, too. The ability to work becomes a requirement of industry and over compensatory productivity. Yet the role of *mother* cannot change--maternal expectations remain intact, creating a dualistic ideal of a career-having supermom that makes killer deals and is still home by three to cut crusts off sandwiches for kids after school. This shift starts to push out the critical balance of mothering and fathering, in which authoritative (not authoritarian) parenting from fathers is slowly crushed as its value is increasingly vaguer in a world of "I can do it all" women.

The loss of mothers in the home through forced dual-income attainment and social expectations to fulfill suffragette dreams of equality is disastrous for children. It's disastrous for women, too, as it removes choice from the ancient human tradition of foundational mother-child bonding and caregiving through infancy and early childhood. During these early years, a period of unbelievable neuronal growth and brain plasticity, core concepts of emotional regulation, empathy, compassion, and a host of attachment behaviors are formed and concretized in the neural circuitry of children. This is not a demand to "put women back in the home!" but rather, a plea to look at the priorities of our society, communities, and families.

This is all within the bell curve of women's rights and supporting women: *choice*. Choice in working a little, working a lot, being home when school is out, or being active in educating your child while they are learning and growing. When the freedom to raise children is removed as a result of financial demand or social shame, we have lost choice. During formative early years for infants and toddlers, the critical foundations of emotional regulation, healthy attachment, and secure identity are profoundly specific to maternal caregiving. Does this mean fathers, grandparents, aunts or uncles, loving foster parents, or older siblings can't provide it? No. It is certainly still possible. It also doesn't mean that all women are stable, excellent mothers. An uncomfortable truth is that maternal instinct does provide a general footing for nurturing new humans in a way that is often measurably superior to other scenarios of child-rearing.

Sitting within and navigating uncomfortable truth is one of the most productive ways in which we can grow as people. As a single mother with an infant and toddler for six years, someone saying my kids were better off with a mom and dad, or at least another parent in their lives, was a gut-punch. *I work so hard. I love them so much. I do everything I can to ensure their health and happiness. What do you mean, I'm not enough?* This discomfort led to real thought about why it was infuriating, why it made me angry. Outside the fact that speaking into someone's life without knowing them well is judgement as welcome as flatulence to the face, there was more to consider.

I thought about how exhausted I was every night. How we couldn't afford gifts or Christmas trees, and how a car breakdown felt like a death sentence. I thought of how ashamed I was to go ask about enforcing child support, and how the shame doubled as I left in tears when they told me my fear over violence or retaliation meant I shouldn't file for it. Then there was little moments of clarity; some were immature and sad, like lying about selling a couch to "open up the space" when we didn't have any food. Others were downright funny, like when my three-year-old son sat on the stairs naked, pointing to his testicles, saying, "What are these?" I had only myself and my daughter—I'd been resolute about real names for body parts and age-appropriate honest information, but I'd completely forgotten to tell him about that basic anatomical part of himself. There were little missing pieces, and through all the hubbub of finishing school, working, and being a fulltime mom, I could not possibly fill them in. Like carrying an armful of laundry, underwear and socks will fall along the way whether you notice or not. The financial, physical, social, emotional, and mental load was heavy, making little space for fun, playful mama. Or, for that matter, fun, playful *woman.* The other missing parent, gone due to exhaustion and feeling overwhelmed, was *me.* I would see friends occasionally grumble about feeling "like I'm a single mom" when their partner or husband was working or traveling, or if they had shared custody. It was difficult to remain compassionate, because the difference between *having* another parent and *not having* another parent is the difference between having food and not having food. Rarely does a starving person complain about the taste.

Having another parent is unbelievably important. This is particularly true as children age. In adolescence, a second period of critical plasticity occurs. During this time, children often gravitate more toward the authoritative parenting of fathers, as these more naturally paternal behaviors encourage independence, critical thought, conflict navigation, and development into young adulthood where identities are all but finalized. We've talked about maternal instinct—yet what gets undersold, underrepresented, and underappreciated is *paternal* instinct. The concept of protection, provision, instruction, and preparation for life with firm control over the direction of the self. Fathers provide something equally as critical as mothers: identity. While a nurturing mother provides emotional regulation, safety, security, and healthy attachment behaviors, a nurturing father provides guidance, accountability, expectation, and challenge. The foundational self is given the opportunity to become the aspirational self.

The Latin word *nutrire* is the root of both *nutrition* and *nurture;* to nourish, feed, or foster. From the specific nutritional needs for healthy development to the broader nurturing needs of compassionate teaching and selfless caregiving, children have a host of *nutrire* domains that require sustenance. Developmentally, a mother that never stops babying a child creates an insufficiently nourished adult, in which a sense of self independent from mother is less clear. A father that doesn't model and engage in clear boundaries and consistent expectations for a teen produces an undernourished young adult without respect for others as a symptom of craving leadership. Two active parents, loving and committed to raising healthy kids, provide an incredible buffet of social-emotional, cognitive, physical, and spiritual nourishment. It really is quite incredible what parents do with the right intentions—despite all of us making mistakes.

Parents don't always stay together today. They don't. We remain at around a 50% divorce rate for first marriages, and 60% or more for second marriages. This could be a host of issues, but primarily, we can see this as an inability to select, maintain, and grow in tandem with a mate through shared values and firm individual identities. It could have been an issue in any of those stages, and all provide ample opportunity for abuse, resentment, unfaithfulness, and brokenness in the children produced during such battles. The freedom to marry anyone and divorce any time for any reason has become, again, a measure of poison in society and for individuals. Summarily, we don't have to think before we act, and we let our genitals, ideologies, and emotions make choices that are unsustainable and harmful; we then have to make ways out of those choices, too. This creates a tragic and extreme pendulum swing between far-flung modes of partnership, from one marriage forever without choice to intimacy without commitment and perversion within and around it. The fact that there are phrases like "minor-attracted person" to replace "pedophile" or "child molester" should tell you, without question, that individual freedom can indeed become poison when it translates to excusing predatory behaviors and failing to protect the most vulnerable among us.

The stress on American families is incredibly deleterious to our health and wellbeing, and much of the stress is derived from governmental and social mores of productivity, taxation, identity politics, and media consumption. If we are not producing something that earns money, fame, or power, it is considered useless. Consider fine art, for example. Creating a work of art to express, communicate, and entertain becomes a hobby undeserving of considerable time in one's workweek, unless that hobby is a "side hustle" that produces monetary value or a trajectory toward monetary value. Now consider parenting. Inarguably more important than painting or self-expression, parenting is the foundation for building healthy children into healthy adults. Parenting is child-rearing, which equates to the transfer of culture, value orientation, worldview, and *identity* to developing human beings, and yet parents who can and do choose to stay home with their children to raise them up well are "stay-at-home" moms and dads, with an underlying value assessment that they aren't contributing financially, so it's less valuable. Maternity leave, on average and if you can access it, is ten weeks in the USA. That means that mothers return to work and put the care of their infant into the hands of others just as their baby is starting to lift its head up and flop over on their back. Paternity leave is even more abysmal, and for the scant few able to take it, it averages only around 17 days. Parents aren't legally entitled to take parental leave--only around 21% of people in the US actually have it as a benefit. Back to work. Be productive. That's your meaning--not raising your children. Your meaning, your identity, needs to be consumable--so unless you're creating a hit YouTube channel that shows the best parts of your day and avoids messes and arguments, you need to get your ass to work.

Adoption and fostering remains another area of stunted parental leave provision. The herculean efforts I observed my own sister take in the adoption of their two children was a reflection of our mother's dedicated approach to raising children. Consuming everything she could about attachment, honoring culture, effectively and sensitively learning and practicing African hair protection and styling, and countless other considerations would have been far less feasible if she hadn't been an early childhood educator and owned her own school. There is no requirement of our employers or society to provide attachment and bonding leave for adoptive and fostering parents. How incredible would that impact be on the children most desperate for it?

Parenting isn't easy. It is a producer of happiness, however, because happiness is a byproduct of meaning. Wired for creativity, learning, exploring, and solving puzzles, our brains thrive when we take part in learning and value transmission. Having someone to help develop into another puzzle-solving, creative humanoid is tremendously meaningful. The joy in it is often hacked away by the stress of living in a consumer culture that says happiness by way of meaning takes too long. You could just buy little bits of it at a time, in whatever dosage and style you prefer. A little fix. But you need the money to buy those solutions. So you need to go to work.

The number one reason for conflict in couples has been, and remains, finances. It's the most argued-about, the most reviled, and the most toxic topic for relationships. Yet cited divorce reasons include lack of commitment, infidelity, lack of family support, and constant conflict as primary drivers for many divorced couples. What do all of those things have in common? A fracturing of identity--unable to fully see oneself in a partner role, seeking missing romance or pleasure from the marriage, lack of familial structures and help, and the hardship of never quite feeling aligned with your partner. Is it any wonder that we fight about money, and divorce over loss of identity? The common denominator is that we are pushed into productivity for society while neglecting our internal selves, and therefore, our interpersonal relationships. Poorly formed individuals without a firm sense of self are vulnerable to all of the aforementioned reasons for divorce, and we're societally stuck in a cycle of poor identity formation because we're never home to parent our children well. We aren't there to nourish their identities, and yet we're mystified that they cleave to influencers and ideologies and victimhood. They have no firm sense of self yet, and the people that help them find it--guide their wagon, if you will--are punching a clock and being a powerhouse career parent; they're rationalizing drinking in the car on the drive home because they know their paycheck isn't enough and they're just going into a hell house of stress and mess and this is the only time they have to be *alone*. We crave peace so much, yet our place of peace that was once our family is mired in financial stress, housing crises, health insurance premiums, grocery costs, metabolic disorders, anxiety, credit scores, and ultimately, a sharp lack of meaning. Remember how meaning produces happiness?

Attachment in early childhood is critically important to form healthy personal schemas and self-regulation of emotions. Stress management is sketched out in infancy through emotional co-regulation, and even tiny toddlers look at caregiver emotions and responses to modulate their own. Toddlers that experience a minor fall will look at their grownups. If the grownup is smiling, clapping, and saying, "You fell, and you're okay! Good job, come here!" the toddler will almost always pick themselves up, try again, and perhaps shed a few tears. A response of rushing toward them to pick them up, panicked and saying, "Oh gosh, baby, are you okay?!" will almost definitely inform the toddler they're *not* okay, and tears ensue despite no injury and the opportunity to overcome the hardship is stolen by the parents. Similarly, simply *pretending* a baby is hurt will make them cry like they actually *are* in pain--tiny humans use their caregivers to help them understand their world, themselves, and the interplay between the two.

Attachment is disrupted with the removal of parents from primary caregiving roles. Maternal stress, which we'll cover more in the developmental impacts of trauma in our fourth chapter, impacts attachment as well. Couple this with financial strain through poverty or divorce, lengthy work hours outside the home for both parents, exorbitant childcare costs, and questionable quality of affordable childcare...you have a recipe for children raised on screens in the company of strangers.

This is the removal of parents through socioeconomics. If you do not wish to work the number of hours at the rate of pay or job you desire, but you *must do it* to ensure your family survives financially, medically, and socially, you are no longer free to create. If a cancer diagnosis bankrupts your family, and if choosing creativity over productivity equals socioeconomic punishment, you are not able to pursue life, liberty, and happiness. If you must give up your parenting time, your generative hobbies and activities, and your limited free time to perform side jobs to subsidize your life...you're no longer an independent American innovator and freeperson. Your access to socioeconomic mobility is increasingly narrower, and having a family? That crowds the path to success and solvency for most American parents even more. Yet we want kids, we want a family—so that narrow path means something has to trail behind and receive less time and attention—and that something can rarely be quitting a job that feeds your family. If your choices are employment and healthcare versus family versus creative pursuits and you can really only do one and a half in your limited hours of the day, you really don't have a choice. The majority of Americans as of 2019 (around 60%) were one missed paycheck from poverty, and 40% are one missed paycheck from homelessness. These are numbers *before* the pandemic and *before* the significant inflation rates as of early 2025. Without that mobility and freedom to exit poverty...you are a forced laborer with citizenship.

These social shifts have enormous impacts over time. Now, the expectation is that both parents work, and if one stays home, it should be mom, because a) she's less valuable in the marketplace and b) dads aren't really that useful in the home. It's the implicit and underlying value assessment that we don't like to talk about, but nonetheless, adhere to. The concept that women *can* work but are *fortunate* to have a husband that pays for things so she doesn't have to quickly translates into *women being equal in theory, but in practice with children they need a man's income.* The idea that men *can* be critical child rearers but are *fortunate* to have a wife that stays home to permit him to work translates into *men are important parents in theory, but in practice with children are really only good for paying the mom to stay home.* If this seems misguided, antiquated, or untrue, let me take you on a brief journey into state-sanctioned removal of fathers from their children's lives.

My husband gave me three wonderful stepchildren, adopted my two babies, and routinely gives himself completely to parenting and provision of all we need. I see our three boys blossom with his teaching, and our two girls operate with safety and security in who they are, where they are going, and their own internal value. He also gave me the sharp gift of his ex-wife, who is necessarily in our lives and often a significant source of stress. If I hadn't lived it for the past several years, and hadn't researched it recently, I would never believe the destruction brought about by skewed and gendered ideals of finances and caregiving. The revelations when I was first learning about all my husband had gone through caused me to investigate other divorces, particularly where the father was mired in false accusations and parental alienation.

I began researching divorce in the US during this eye-opening experience in our first few years of marriage. I started asking fathers across the country about parenting, about their kids, about their treatment in the civil process in their home states. These fathers had *a lot* to share about what they went through in their divorce, and the abysmal representation, adjudication, and outright judicial favoritism toward mothers. I began researching existing studies--is this true? Does this happen like it happened to us and so many of the men I learned from? It turns out, yes, yes it does. Despite many states claiming equitable parenting as a default mechanism for parenting plans and dissolutions with children, bias toward mothers permeated every experience. Typical responses from fathers to my questions included sharing that they were rigorously investigated financially, while the mothers were rarely investigated with that level of scrutiny for unclaimed income. Fathers were routinely forced from the family home to decrease conflict, and then immediately labeled as abandoning the children or as a secondary caregiver only. Mothers were perceived as gentle and nurturing while fathers were controlling and manipulative, despite overwhelming evidence that the marital conflict was mutual and often instigated by the mother. Research on high conflict divorces revealed that implicit biases regarding attractiveness, soft tone of voice, demur or vulnerable presentation, and other feminized aspects of mothers informed custody decision-making, despite both parties being capable, loving parents. Fathers pleading to be with their children were viewed less favorably than mothers.

Cases of mothers pursuing criminal charges during high-conflict divorce cases were common. In a class designed specifically for law enforcement officers, I asked two groups to read the same phone call transcript to 911--one read it as an upset woman, and one read it as an upset man. Across the classroom of officers, criminal charges were assigned to the father from the mother's call, while the father's call was deemed civil, non-criminal, and a matter for the court. *It was the same transcript*. This has a number of reasons behind it, some valid, and some less so. Officer discretion, a huge part of determining risk and lawfulness, reasonably places officers in a position of erring on the side of caution. The issue is that the leveraging of *potential risk* becomes a powerful tool for manipulating gender perceptions, often at the cost of someone else's reputation, and in these scenarios, being with their children.

The financial impacts of divorce remain horrific. Alimony, disproportionately from ex-husbands to ex-wives, is an extortive practice leveraged against wage-earners enduring divorce. It is still commonplace, and without consistent, logical, or equitable application. Child support is also a broken formula. It's based on parental income, not relative need, making it punitive for high wage earners. It also encourages underemployment for those capable of working that don't want to lose the higher rate of pay in support monies by increasing their own taxable income. Disability income is included in support calculations--multiple fathers I researched were significantly or totally disabled, required to pay significant child support, and were routinely experiencing their children being withheld by the mother at her whim, against Court orders. Child support payments are not taxable, either--so divorced mothers once more are encouraged to maximize support income and minimize independent income. Again and again, fathers would state that they only want to be with their kids. They just want to raise them. They just want a fair shake. Again and again, they are told by our civil processes and judgements that they are unimportant in their children's lives--following parenting plans is civil, but failure to pay support very quickly becomes felonious. Even filing contempt or requesting the parenting plan be enforced can range from filing fees to attorney retainers and more—yet the effectiveness is negligible. The "primary parent" has all the power to poison, withhold, delay, and lie. All while the other parent tries to navigate the costs of more legal fees, more attorney retainers, more evidence collecting, more conflict with the other parent, potentially more allegation from that parent, and the cost of their own housing and daily activities. The highest cost is the inconsistent, lost time raising their child. The most salient theme from divorced dads was heartache that separating from their spouse was viewed as separating from their child—but it was simply untrue. Many reported that the conflict and trying to have a voice and identity as a parent seemed hopeless, and that maybe just giving up would be less stressful for their kid.

Absent fathers are not always absent by choice. It is necessary to determine if their absence is not disinterest, but due to constant conflict with mothers that often results in biased criminal charges; financial devastation preventing consistent parenting time; parental alienation by mothers to support their identity as sole-source parents; or the insurmountable identity crisis that compounds an existing fear: they aren't important in raising their kids.

This is not the majority of divorces, but echoes of these issues are within many separations with children. This is why, when examining childhood adversity (also called Adverse Childhood Experiences, or ACEs), divorce is listed as a traumatic event. Now throw in the issues of scarcity, pervasive and generational poverty, parental incarceration, substance misuse and dependence, increased risk of sexual abuse, increased risk of physical abuse and neglect, psychological or emotional abuse, and a climbing likelihood of experiencing or witnessing interpersonal violence...you get the point. Parents forced out leaves a vacuum in child development, and it gets filled up real quick with all sorts of little poisons masquerading as coping mechanisms.

2

CREATE VICTIMS, REAL OR IMAGINED

The United States has always had a little bit of an identity crisis. A newborn with anger and teeth, we nationally struggle with the massive size, scope, and complexity of cultural mores, value sets, and ultimately, who we believe we are as a nation. American ideals of individuality, independence, innovation, and resilience are slowly supplanted with individual narcissism, loneliness, overwork, and exhausted atrophy. We remain, unequivocally, a powerful example of doing things our way and fighting to our last breath, but beneath that macro level, in everyday life, we are individuals largely at odds with ourselves and our nation. We've uncovered massive amounts of both outright and insidious manipulation in our politics, divisive group-think perpetuated by the corporate media, and the rates of dishealth in our children to our aging adult populations are increasingly alarming. Diminished numbers of military-eligible youth and young people, metabolically disordered citizens encountering disease younger and younger, and an overdependence upon quick-fixes and interventionist care have taken our country viciously by the wrist and forced us into a corporate system in which we are inputs, outputs, and profit margins.

A key factor in making sure that people buy cures is to never stop selling illness, and to hide truly healthy products, behaviors, and solutions. For example, diabetes is the primary cause of chronic kidney disease (CKD), along with obesity-related issues including but not limited to heart disease and long term use of commonly consumed NSAIDs and drugs like lithium. Designed for individuals with CKD, *Novasource* is for "people with chronic kidney disease (CKD) on dialysis, acute kidney injury (AKI), fluid restrictions due to CKD or AKI, or electrolyte restrictions. Increased protein content helps maintain lean muscle". Obesity, heart disease, and diabetes are all metabolic disorders (specifically cardio-metabolic for heart disease), and the number one factor in those metabolic conditions? The consumption of ultra-processed foods leading to obesity, insulin resistance, and inactivity. It would be really lovely if the altruistic makers of Novasource (or Isosource), or perhaps *Compat,* the medical equipment provider, could convince massive corporations making such unhealthy, metabolism-destroying products to stop doing so. It would be nice, except Nestle, responsible for a massive number of ultra-processed food products, owns Novasource, Isosource, and Compat. They own the medical equipment and disease-oriented supplements needed for you after you over consume their other products. It's a shockingly sustainable model if you simply don't care about people.

There is a disgusting logic you can apply in this way to a number of different issues in American culture. Obesity is, by far, the best way to get customers. It's even more profitable if you make sure society celebrates you for being obese through "body positivity", while also selling impossible bodily perfection through aggressively filtered hyper-athletic or rail-thin, perfect-skinned models with zero emphasis on the reality behind apparent perfection. Commercials with elite athletes drinking Pepsi, happy thin hikers pounding down Mountain Dew, and skeletal models demurely eating Burger King are advertising ultra-processed food through energetic sexy imagery. The dark irony is that consistently consuming those foods will one hundred percent cause metabolic disruption, disease, and early death. Less sexy.

This formula extends far deeper than appearance and metabolic issues. It reaches deep within our psyches, like dark water seeping into hairline fractures in our development to fill in the cavities left by absent

So here we are, with generations of kids moving through deficient, often political public schooling and social media consumption without parents healthy and available to teach about how the ideas and concepts kids see are managed and understood. Children are being given identities not through cultural transmission, family tradition, community engagement, and meaningful dialogue...but through changing and conflicting ideological concepts that are largely generated from corporations profiting from fear, confusion, disease, and discomfort. I do not believe that CEOs are sitting in fancy boardrooms plotting the destruction of American youth and families--yet intentional or not, *this is the outcome.* Paying politicians to support the incarceration of low-level non-violent drug crimes to fill private prisons is deeply immoral. This is no different than funding the FDA to fast-track unsafe or unhealthy products, or spreading fictions about effective therapies that do not provide high returns to investors to prevent their approval. Whether intentionally destructive or simply "doing business", the outcome of making people hurt and then offering to help them feel better is, at its core, pathological.

Now the call to "being someone who matters" is rooted in victimhood--even college entrance essays ask about hardships experienced in life as distinguishing elements to the application process. Calling upon adversity to give personal meaning is different than calling upon *a healthy response to adversity.* Celebrating what you have intentionally done, rather than celebrating something that has happened to you, is a critical step in resilience and antifragility. Children raised by a society rooted in commodification and marketing of ideas and products need parents and caregivers that are able to block nonsense, teach navigation of hardship, and provide consistency in order to have any chance at an identity that comes from within. That small period of early childhood with squishy neural development and plasticity needs family, not TikTok. Consumerism is a symptom of abuse--at societal and individual levels. It is a desire to fill up empty spaces that should have been full of safety, love, and an identity that knows it matters. If you are consumable, you don't matter. You're just a project, a thing, for use by outside forces. Children without a firm sense of self due to neglect, abuse, and consumer culture are children who are fundamentally dysregulated, stressed, and will seek external controls over internal feelings. In the first chapter, we talked about the removal of parents through socioeconomic pressures; we talked about the roles of parents and the importance of attachment and effective child-rearing. This is where we end up when we remove the guidance from the home and leave children to their own, very literally, devices. When we examined European student hopes and dreams, they were largely intrinsically motivated aspirations; American students' dreams really didn't hit that mark. The removal of parental influence is dramatic—and whether it's through generational trauma, poverty, time scarcity, or social media, its parents that are losing their role as gatekeeper for their children.

It's little wonder that while European countries all have differing values and structures, sizes and opportunities, they refuse to prioritize the economy of their countries over the health of their families. In the USA, we reward excessive work hours, while many European countries cap *overtime* at 43 hours a week. Parental leave in Europe is generous—eighteen months in Sweden—and universal healthcare means that employees are not bound to employers to maintain healthcare. Americans deserve these humane opportunities as part-and-parcel of our citizenship. I love this country with my heart and soul. It is the escape from forced labor, illness, theocratic and ideological governments of oppression, and the pursuit of individual freedom that makes us, us. If those things aren't true, who are we?

3

SELF-EFFICACY MAKES MEANING

Self-efficacy is a beautiful construct of having a firm locus of control within your life. It means that you believe you are capable, and that things you need and want are possible to attain because you have the skills to do so, or to learn to do so. It is the infant that grows into confidence in youth, direction in young adulthood, success in life, and a sense of meaning and purpose through each stage. Self-efficacy is also one of the first heads to be lopped off with adversity in childhood, particularly for victims of abuse, neglect, and violence. Consider a child who is routinely beaten by his father, the age-old story of the alcoholic dad who comes home looking to beat on someone, and his son doesn't want his mom hit again so he provokes the drunken bear into focusing on him instead. What is this child's sense of self? Of control in his life, of destiny, or hope? This child is growing up learning that stronger, meaner men hurt people who aren't equally mean or strong. That being vulnerable means being victimized, and that the decisions and liberties within the home belong to the one who can cause the most harm. How will this inform his future relationships? How does this show up in school, when someone tells him he's weak, and looms over him, and suddenly he can't see or hear but he's responding to that kid like he's his dad--a threat, and he has to be stronger. Now he starts shoving other kids, putting the beat on them just to make sure they know he's the stronger, meaner one. It's a simple and logical path from being abused to abusing others, because it's the template you are given and live in every day.

Compulsive disorders are an interesting expression of desiring control over one's life, as they are largely rooted in trying to manage unknown, unpredictable, and uncontrollable events long gone but manifesting in different ways. The most common history of patients with anorexia nervosa includes chronic sexual abuse. Can you imagine how a brain might move from "I have to be used up in secret for someone else's pleasure, and I have no control over what they do to my body, and nowhere to turn, and no way to stop it" to "I can control my eating"? Add in body dysphoria, interrupted pleasure and inhibition management, and a trained understanding that being hungry is a sign of control, and this individual who doesn't eat is doing so because their brain feels life is controllable when they are hungry, because it's in their *choice* to stay hungry or not. When we have all our power taken from us, we seek it in any and every place possible.

This balance of inhibition and motivation (pain, pleasure pursuit) is seated in the *nucleus accumbens,* and it is responsible for maintaining homeostasis, or balance, between doing and not doing. For people struggling with compulsive thoughts or behaviors, completing the task is critical to being able to think about other things--it's like a fix during withdrawal. I noted this during a time I provided case management for individuals with chronic health conditions. Often, being chronically ill, or experiencing consistent levels of pain without reprieve, can create a slow-burning toxic stress in which recovery from illness seems impossible.

A woman I worked with had a history of abuse as a child, and was clear about connecting compulsive hand washing with being slapped and burned when she was little for minor dirtiness or unkempt appearance. This was exacted upon her by a stepmother displaying aggressive jealousy of her, taking over hygiene policing in the home, though targeted at her young stepdaughter. As a teen, she would wash to the point of chapped, bleeding hands, and felt that if she didn't wash them, she couldn't even drive her car because the need to wash her hands was too strong to concentrate on driving. During therapy as a young adult, she recognized that her recovery was rooted in the understanding that being abused as a kid made her obsessed with being clean, and that not washing her hands created feelings of filth and disgust. "Washing my hands is how I cleaned my soul," she said. "I couldn't control being hit, or when I'd get hit, but I could wash. I kept thinking that I had germs on me, and I'd get worried my stepmom would see it, or sense it." She would also obsessively clean her hands before interviews, dates, or important meetings, because she could go only an hour or so before she started thinking about where she could wash her hands again. "It was like being an alcoholic, like thinking about where the nearest bar or liquor store is, except I was thinking about sinks and soap." To concentrate, she had to make her brain feel like she was the driver, and that meant washing up to stop thinking about washing up. Self-efficacy for her was taken when she had her safety and acceptance removed through abuse, and her brain did what brains do best: use the path you have. You'll be ok if you can just do your routine, just do the thing, just wash your hands, and I'll flood with dopamine relief, and make it even more likely that you'll keep doing it because of that dopaminergic cycle. Washing hands is a good thing. Doing it until your hands bleed, you're late for work, or you are compelled to leave a date early is maladaptive, and doesn't make you healthier, happier, or in control. I believe this woman had other issues contributing to her hand-washing; it wasn't purely a sense of dirtiness from childhood. It was far more likely it was the culmination of many uncontrollable things manifesting in hand-washing.

A lack of self-efficacy shares many features with a lack of identity. While identity is a sense of self, efficacy is a sense that the internal self can do things well. Identity and self-efficacy intertwine, and often these interconnected features are a result of lifestyle feedback loops, like being a cop or in the military. It also extends to believing you are sick, or victimized, or cannot overcome chronic behaviors because it's "just how you are". An area of compulsiveness is often separated fully from it, but nonetheless belongs firmly within it: addiction. Addiction is a sneaky, all-encompassing capture of an individual that drives them to consume, engage in, or ruminate on a substance, act, or belief. Addiction is imbalance and dysregulation: your brain activity that motivates and inhibits is swinging to the extreme, as feeding addictions creates a dump of happy chemicals, only to be drained in equal measure once the act is done. It's like trying to go for a walk, but it's like walking on a soundwave that is constantly undulating. You just want to balance, but you're always going up or down, getting back up when you're too far down and careening down when you're too high up. On a much less extreme note, think of when you eat a lot of Chinese takeout, and feel stuffed, only to be really hungry again in thirty minutes. That's because the carbohydrates in the rice and sticky sauces have spiked your blood sugar, and the crash immediately makes your body beg for quick fuel again. The constant back and forth is due to inordinate levels on one side, requiring equal and opposite response.

Rumination, a key component of depression and obsessive-compulsive issues, is remaining stuck on a thought pattern or concept in such a way that interrupts your wellbeing. Rumination is also a part of your memory system, and exploration of being unable to control one's thoughts and memory recall leads very quickly to the realm of post-traumatic stress disorder (PTSD). This is precisely why gratitude journals, talking to trusted friends, and meditation are so helpful for less severe issues of rumination. They are strategies that support healthy recognition, identification, processing, and release of thoughts and feelings that are interrupting your ability to look at the world through a lens other than what is causing you pain. Research on gratitude as a practice reveals, especially for adolescent girls, improved self-worth, confidence, ability to navigate stressors, and a view of the world that is significantly more positive. For girls during adolescence, internalized dialogue and negative thoughts are rampant, and worsen with obsession of the deluge of social media demanding constant attention to identity and sex appeal. Practiced gratitude for positive experiences and individual worth creates habits that inform lifelong behaviors and enhanced self-efficacy.

Used fast, loose, and often, PTSD actually has quite a few diagnostic indicators that move it beyond "feeling triggered" to "I live in fear of my memories, and my memories live outside of time." While many behavioral health advocates insist on monitoring or policing overuse or misuse of mental health terms, I lean toward being reasonable and respectful. Saying, "I have such PTSD from that phone call" doesn't necessarily equate to believing a difficult conversation is synonymous with witnessing a murder. And yet, importantly, understanding that truly qualifying PTSD symptoms can absolutely destroy the lives of people is essential. Painful, persistent memory recall is a hallmark of PTSD, and includes symptoms ranging from rumination and compulsive thinking to being thrown back through time to the trauma event through somatic responses to environmental triggers. Complex PTSD presents yet another challenge for individuals and their clinicians to manage--it means a host of symptoms, levels of severity, triggering events, and dysregulation all require nuanced and complex approaches to treatment that can be disrupted at any time, making consistency in treatment all but impossible. Complex PTSD is often found in circumstances of extreme poverty, war, and polyvictimization. Veterans often come back from deployment with complex PTSD, particularly if they experience multiple combat deployments. While it is possible that soldiers not experiencing battle have PTSD from service, the engagement in life-threatening situations for sustained and acute periods is a crash course in brain adaptation and survival. In World War I, for example, soldiers encountered mechanized weaponry at large scale for the first time, and the impact was devastating physically and psychologically. Mired in filthy, muddy trenches for *years*--that's right, years--meant that disease, grotesque violence, completely decimated bodies, and consistent loss of brothers-in-arms were daily events for sustained periods of time. Men that had "shell shock" were soldiers so unbelievably traumatized by this horror that they'd soil their pants, become catatonic, have psychotic breaks, and often become institutionalized or die by suicide. This was met by the public, unaccustomed to warfare and blithely unaware of the horrific violence and trauma experienced by these men, with at best pity, and often, shame and derision. Called "psychologically weak" and "mentally deficient", it was decades before PTSD was identified, labeled, and treatment options offered. It remains an issue of shame *still*--admitting war and violence has had a lasting and negative impact is still an issue of confusion, shame, and avoidance.

While the cause of PTSD in Veterans is, largely, war and all its trappings, there are resiliency measures within brotherhood and tribalism that speak to a shared warrior spirit. In tribal societies, war was shared amongst the group with a general understanding of impact. In massive, modern warfare, it is select warriors who fight fully separated from society, which may or may not approve of the cause, circumstance, or entire war itself. Even if celebrated as heroes, returning soldiers lose the primary protective factor available to tribal warriors: they lose their brotherhood. Repeatedly, the identity formed through tremendous adversity and deep, life-altering loyalty in a shared combat context is lost upon return to civilian life. The sense of purpose on missions, the sense of profound trust amongst brothers, and the self-efficacy in making the cut, training successfully, and executing tasks with violent precision is far, far more difficult to mimic in everyday life. Inserting a warrior into a grocery store or play performance is a context in which their skillset is not readily or easily attuned. I have, with considerable amusement, gently reminded my husband that casually saying "fuck" twice in parent-teacher conferences may change the dynamic differently than with his Ranger buddies over a beer. War is a context in which aggression, decisiveness, intense loyalty, clear purpose, and hyper accountability are not just wanted, but fundamentally *required* for survival. Imagine, then, coming back to a society in which your aggression is viewed as threatening, your decisiveness is controlling, your loyalty is never met with equal commitment, you no longer receive direct orders or missions, and your expectation of accountability is abysmally met by regular people who take sick days and don't always call when they're late. You no longer fit into the world, and your identity is fracturing painfully.

The ability to manage pain, violence, and experiences that produce overwhelming levels of stress requires internal and external scaffolding. Internally, the ability to self-manage emotional response, process and respond to stress effectively and productively, and create critically constructive methods of management and coping are essential. Externally, support structures through trust-based relationships, tribal loyalty and identity, and acceptance of the impact violence has on the psyche with appropriate support through empathy, compassion, and social processing helps manage events that *could be* traumatic. These internal and external structures of processing and community diminish the physiological impacts of such acute and potentially toxic stress events.

Combat Veterans don't have that. The lack of widespread understanding, empathy, even compassion is a significant hardship for soldiers returning from a context where violence and their response to violence are foreign to larger society. They separate from fellow service members, observe and interact with media pundits claiming hollow celebration or outright disgust at their actions under order, and encounter conflict with interpersonal relationships. This leads to real, perceived, and often chronic feelings of isolation. Interestingly, when you examine data on Veterans and PTSD, those that remain in high-performing combat situations, whether private contractors or as part of the military, report *fewer* issues with PTSD and issues of stress response. These elite soldiers have something that returning Veterans do not--warrior tribalism, consistent purpose, and even when back to civilian life, their identity remains intact. It is the fracturing of self through poor acclimation that compounds and exacerbates symptoms of trauma--a dissociation between the self and that self's place in the world. Recall the positive combat traits of aggression and violence of action. Now place that hyper-response to threats in a Walmart parking lot when someone almost runs over your kid through inattentive driving. While quick response was warranted and needed, the ensuing forced removal of the driver through their window was a facet of response ill-suited to civilian settings. Summarily…Veterans are trained to adapt to war and never trained how to re-adapt to society. This creates internal, external, and often legal consequences.

An inability to adapt to new environments is a feature of trauma symptom sequelae. The internal schema for self and response to place, time, and people are altered with situations of significant stress. While the chronic nature of stress through scarcity, familial substance abuse, and parental incarceration creates a slow drip of excessive cortisol augmenting neurobiology over time, massive temporary hits of stress in rapid succession through life-threatening situations can create a more sudden interruption in regulatory functioning. In both scenarios, *chronic and sustained stress* and *acute and toxic stress* do similar things: they alter how you respond to stress events in life. Once altered, there is a restructuring that is required to regulate responses back to normative levels, and for trauma-induced chronic disorders, this is not just a heavy lift. It's an impossible scenario in which the systems required for change (memory, emotional down regulation, critical thought, organization, all the elements of *consistency*) are the very ones damaged by trauma. It's a catch-22, and it very quickly compounds into hopelessness. To get out of the pit, you have to climb, but the footholds are gone and the handholds crumble when you use them.

There is a reason that Veterans, in significant numbers, do not identify traditional therapies as effective. Increasingly, choosing complementary and alternative medicine (CAM) is of interest, with many Veterans echoing what others with chronic behavioral health issues state. This is, largely, that talking to someone who hasn't experienced this level of trauma, plus medication that numbs at best and addicts at worst, is just a pause between symptoms and a Band-Aid on a bullet hole. It's not worth it.

Yet Veterans able to access psychedelic therapy have dramatically different outcomes than those who do not. In many cases, Veterans stated that traditional therapies were ineffective due to the provider's limited experience and understanding of the impacts of combat engagement; lack of interest or access to consistent and ongoing therapy; complexity of symptoms presenting obstacles to adhering to treatment plans, and more. Symptom complexity is an area in which triggering events and loss of support structures due to addiction and other behavioral health issues dismantle and undercut recovery efforts. Addressing flash-bang memory recall through somatic therapy is a typically helpful approach to managing PTSD symptoms. Imagine how difficult it would be to adhere to treatment plans when your symptoms are also causing marital issues, or your drinking impacts your medication routine--again, the pit is deep, and dark, and the way out is broken.

There is, across multiple studies spanning what is now almost decades, an abundance of research into the changes experienced by Veterans following clinically-supervised or shamanistic psychedelic treatment. Most prescription medication, even with cognitive-behavioral therapy (CBT) and other combined traditional therapies, boast a scant 20% effectiveness rate. It's important to note that "effective" is a moving target. Many metrics define "effective" as having some undefined, general positive impact. This could be a reduction in the number of symptoms and behaviors checked off on diagnostic tools, or it could be self-reported outcomes from patients, or a combination of assessment types that attempt to define improvement in symptomatology. The most prolifically prescribed medications (despite, again, questionable positive outcomes), are SSRIs, used for anxiety, depression, and other aspects of PTSD. For more severe and ongoing symptoms--and chronic, treatment resistant PTSD, major depression, anxiety, compulsive disorders, and substance dependence are largely chronic and severe for those with trauma--patients may be prescribed SSRIs for *years.* Limited, questionable efficacy, poorly defined measures of effectiveness, and unmitigated symptoms for YEARS. It's no wonder Veterans with chronic trauma response issues lose everything, including hope, as they struggle through constant painful memory recall, startle responses, addiction, romantic conflict, poor decision-making, volatile behavior, and the slew of other hardships present in their daily lives.

And yet.

We'll talk about this more--a lot more--in chapter eight. But for now, let me just share a miracle with you. A miracle that is non-addictive, all but impossible to overdose on, organic and natural, used in antiquity since time immemorial, and requires an average of 1-4 sessions to provide full efficacy. This miracle contributes to neurogenesis, massively, and is almost free to create. The kicker is this: across multiple studies, an average of 80% Veterans who accessed this miracle as a formal treatment *no longer met criteria for their chronic disorder diagnosis.*

It's psilocybin, a funky little fungus. A dirt miracle.

It's wildly effective.

And in the US, it's not only inaccessible for many Veterans. It's illegal.

4

CHRONIC, TREATMENT-RESISTANT DISEASE

You may have heard, read, or said before: "hurt people, hurt people." This is a poignant statement, and I often combine or interchange it with another: "abusers were abused themselves. Not everyone abused becomes an abuser." The vast majority of individuals who exact harm on others were harmed themselves early in life. This cyclical nature of victimization and abuse is particularly salient in chronic, sexual harm on children. The multifaceted impact of routine trespass on one's intimate personal self is typically marked with deep manipulation and fracturing of identity as most abusers are related or well known to their victims. The impact of consistent violation and secrecy on girls and boys shows marginal differences between the genders. Often guilt, shame, and internalized behaviors like rumination and thoughts of worthlessness are associated with abused girls; externalized behaviors like overt aggression, anger, violence, and oppositional defiance are associated with boys. Both share self-medication pursuits: through dopaminergic see-sawing in drug use, risky sexual behaviors, unhealthy and abusive relationships, and overall anxiety, depression, and compulsion.

It isn't difficult to see the pattern of dysregulation in people who have experienced or are experiencing abuse, neglect, and other forms of significant adversity. The human drive for belonging is one of many root behaviors that is disrupted through adversity experienced in isolation:

belonging means your needs are met, and you are safe. The experience of scarcity through poverty, neglect, and deprivation dysregulates a sense of belonging to the world and safety in your home group. Experiencing poverty from birth, for example, poses different stressors for a developing child than experiencing sudden famine through war or environmental chaos (the potato famine, for example). Both are stressors, and both pose risk for maladaptive response, but the acuity felt in moving from *having* to *not having* creates a different perception of the stability of the world and one's place within it. People accustomed to being poor are far less likely to become deeply anxious and panicked when food is scarce or housing is threatened. Those who have had, can lose. Those who have never had, lose less. It's like *Always Sunny in Philadelphia.* Old poor, not new poor. You know you'll survive, and have ways of doing so. Whether positive or negative, the behaviors you develop to respond to crisis or stress events have the ability, based upon your perception of their weight, to support or diminish future stress event management.

The issue lies, of course, in whether the adaptation to the stress event persists well after the stress event is gone. Like the bear in the road, like the war in the mind, surviving adversity requires management of your own response, and for cumulative, chronic, acute, and severe stress events...you lose the management role and become the passenger in your own survival. Even if it hurts you--even if you observe it injuring your life and wellbeing, you don't have control over it.

Hurt people, hurt people. Much of this is all derived from the same root events: when a person's safety and wellbeing is threatened, injured, and abused, the measures of control to mitigate the harm are removed from the individual. This is the seat of victimization: you don't have any say in what happens to you, and you have no value outside of what can be extracted. This is typified in compulsive disorders, where any measure of control feels like medicine. Whether you're controlling your food intake, your romantic partner, your business deals, your finances, your drug intake, or how often you check a light switch...whatever it is, needing to control to the degree in which other inputs do not matter is an over compensatory behavior mitigating a loss of control earlier in life. Consider also the impact of an unstable or mentally ill parent--never being capable of pleasing them as a child easily turns into an adult lacking distinct identity. Fast-forward to adult relationships and you have the ingredients for codependency, victimization, and extortive dynamics.

This is the increasingly annoying issue I have with *diagnosis.* While a useful tool in identifying specific issues, from an epidemiological perspective, diagnosis is challenging when it comes to complex trauma typology and expression. Trauma is *often* complex, as one event or circumstance can lead exponentially easier to the next. This is the nature of chronic, treatment resistant disease: like a snowball rolling downhill, it gathers volume and nuance contributing to the complexity and symptom sequelae that make it even more challenging to stop or augment. In managing chronic medical issues, co-occurrence of other disease and disorder is common--poor diet and limited physical activity leads to diabetes leads to neuropathy leads to injury leads to infection leads to overwhelmed immune systems leads to...you get the picture. We lump more and more diagnostics on symptom sequelae, when in reality, a break in the cycle is needed. In the example of poor diet and diabetes, treatment in the USA often includes long term *management* of chronic conditions. A solution may be found in readjusting metabolic health: eat less. Move more. Slow steps. But that's not sexy, not easy, and it's not immediate--so we see a rise in "other solutions" that are more immediate--things like Ozempic or other proponents of false satiety signals that again address *symptoms*, not *causes.* Metabolic disorders absolutely exist--and they are largely, if not almost entirely, due to contributing underlying factors that are changeable with prevention (positive habits in a loving and healthy environment), mitigation and maintenance (excellent nutrition and exercise), and honesty (truly controlling what you can, and knowing when to ask for just enough help to make you capable).

When addressing the complexity of trauma, use of diagnosis can run rampant. Many children and youth, when assessed for ADD/ADHD, also meet criteria for anxiety and depression. Depending on your clinician or physician, pinpointing one or the other pushes the patient on a trajectory that may not include the variety of other issues inherent to their particular, individual situation. If medicated for attention issues and still presenting with anxiety, clinicians are faced with a crossroads question: change the diagnosis? Add one? Medicate both? What about when they return in a year, stating they're depressed, too? We develop critical paths forward based upon symptoms, because we like little puzzles we can solve.

Unlike little puzzles, humans are far, far more complex. We are constantly churning through our perceptions of the world and our place within it--taking in data and processing through our particular worldview, self-augmenting and concretizing aspects of our operating system. When your internal operating system is programmed to respond to stress in ways that initially protected you, changing those responses to stress becomes increasingly difficult to do. Healthy patterns and habits are those that enhance your quality of life; destructive ones are those that you engage in knowingly or unknowingly harming your ability to interact with the world successfully.

It's perhaps useful to use a comparative that is outside the broken approach to healthcare in our society. We have a markedly different approach to business, for example. In business, if a practice is diminishing returns, lowering perception of product quality, and impacting investments, we hold shareholder meetings. We look at our funding streams and determine which are unbeneficial, unsustainable, and ultimately detract from the business as a whole. We rarely do this with our daily lives--our habits and internalized perceptions are almost never up for discussion, debate, or recalibration until they pose a glaringly obvious risk. Even then, they aren't always addressed. Complex trauma leads to a host of symptoms that compound poor health and contribute to additional disorders and disease, yet we play whack-a-mole with symptoms through largely singular approaches. Combined therapies, like CBT and medication, are improvements but still lack the recognition and treatment of root cause: disrupted self-regulatory processes. This is not necessarily anyone's fault, and certainly not a judgement upon individuals with complex trauma and related destructive behaviors: it certainly is a rallying cry to do as CEOs would do: let's hold a meeting. Let's look at what is hurting us. If it's hijacked systems of response and regulatory function, let's address it.

The issue with creating ample opportunity for diagnosis is the reinforcement that external forces will fix our internal hurts. This is patently untrue. Having a nice house while living paycheck to paycheck and constantly worrying about the mortgage isn't actual wealth. It's the appearance of wealth. Similarly, having a diagnosis to wear as an identity isn't resolving the daily impact of symptom expression on your life--it's a false sense of control over your health. Externally provided solutions to internally held problems do not fix behavioral health disorders. This is why addiction recovery is marked with statements like, "they have to want it for themselves". An example that immediately comes to mind is a young woman at the detox center I worked for that was easily my age, 22 or so at the time. She was beautiful, originally. But years of methamphetamine abuse and dependence drove her to increasingly harmful behaviors: prostitution, continued dependence, homelessness. Bigger than all of this were her four children, all removed from her care. She was supposed to go breastfeed her newborn infant, still in the hospital detoxing from her meth use while in utero, when I noted the car picking her up for transport wasn't the car we'd called. It was another guy, one of many capitalizing on her remaining beauty, vulnerability, addiction, and poor self-worth.

"Please don't go with him. Please. You said you wanted your kids back, and you want your baby. You have your baby in the hospital, please don't go with him." I couldn't comprehend the pull to that shitty Chevy Impala driven by an asshole and leading nowhere good--I begged her. She just sighed, and looked exhausted.

"Yeah," she said. "I just want meth more than my baby. Sorry."

I never saw her again. It put a little crack in my heart that stayed for a long, long time. I tried to think of what could be so aggressively captivating that you'd hurt and abandon your babies for it, and of course the natural answer is, meth is fucking awful. It truly does replace all your normative motivations, replacing things like maternal instinct for needing it. Replacing hunger, replacing love, replacing self-preservation, all of it--meth was the new answer for the addicted person.

It doesn't *start* that way, though. It simply doesn't. Despite the vitriol spewed toward addicts, zero percent of them were outside playing as a ten year old saying, "gee, I hope I screw up my life so much that people hate me and I die early." All children have a singular desire, whether they are in a healthy home or a shithole where they're smacked for standing too close: I want to be loved. Love, in its many iterations, manifests as being important enough for someone to care about you. That's precisely why early childhood adversity is so significant and deleterious to health, and why interpersonal abuse and neglect is so much more harmful than generalized hardship.

Whether you believe firmly in *tabula rasa,* the concept that all people are born blank slates and developed through learning and environment, or if you believe our memories carry through genetics and time from ancestors long past, it remains that who you are in the world is impacted by the world you reside within. Like everything I've explored to death thus far, there is an unavoidable concept of "it depends" and "somewhere in the middle" to this debate. Largely, the "it depends" statement in human development boils down to one beautiful word: epigenetics. The scientific term for *it depends*. At least in my mind.

The commonly held definition of epigenetics is essentially the study of how environmental factors change the expression, but not the code, of genes. Identical twins, for example, may share the same DNA, but express different behaviors, personalities, and proclivities. These changes are within the *epigenome*, and may or may not hold true over time. The variety of ways that we *become* are astounding, as a shared experience of stress between individuals will be wildly varied in scope and impact between each individual, right down to genetic expression.

Throughout this book, I've interchangeably used "stress", "toxic stress", "chronic stress", and "acute trauma", etc. When using the word *trauma*, I am referring to stress that changes the individual's subsequent responses to stress events. Individual stress response is deeply idiosyncratic, and based upon a host of influential factors. Recall "new" and "old" poor. The stress response to scarcity is spiked for those unaccustomed to it, while the normative levels of stress in managing chronic scarcity make it less impactful to have very little. It doesn't mean that the person accustomed to the stress is somehow healthier--the adaptive properties to addressing scarcity may be very, very maladaptive in times of plenty. Consider a foster child, removed for abuse and neglect at four years old, and placed in a home with healthy parenting, consistent meals, and routine grocery shops. This child is accustomed to undernourishment and sporadic periods of food provision--you may find them in the pantry at night, eating through packaging, and hiding food under their bed. In times of deprivation, this behavior is life-saving. The food may not come back. Translated to a healthy and consistently nourishing environment, this is at best odd and almost certainly alarming and heartbreaking. "You don't need to save that, I promise, we'll have dinner tonight," might be a common phrase until the child learns to trust that yes, there will be food again.

We will talk more about this case later on, but for now, consider briefly for context a child routinely told she is not permitted to dress, speak, play, or enjoy activities without her mother. When she does, her mother becomes incredibly hurt, screaming that her daughter's individual place in life is tethered to her, and that she is incapable and disallowed from independent happiness and identity. When this girl grows up, she binges on alcohol and food, forces herself to vomit, selects mates that are controlling and codependent, and seems to have strange and persistent issues with her emotions overwhelming her. She feels shame all the time. The behavior of hiding herself to make others feel better persists. It's also why people from abusive homes often don't even register unhealthy flags in relationships--they are so accustomed to it that their learned behaviors within those dynamics are at direct odds with healthy relationships--so they seek similarly dysfunctional ones. The discomfort unhealthy people feel in a relationship to healthy people often drives them to destroy the health of the relationship in order to create comfort in dysfunction. Working through the discomfort requires a recalibration of what feels normative--which means identifying what is *not* normative in the self is critical. It's also incredibly, immeasurably, and deeply difficult to do.

The perceptual nature of stress is beyond significant in understanding cause, outcome, and solutions. Trauma, the outcome of toxic stress, lives within the perception of the individual--who may or may not register qualified events as traumatic. This is the tricky, hidden nature of trauma and its impacts: they present often without warning, introspection, or control. When trying to lose weight years back, I was reducing carbohydrates primarily, and this meant that half and half was a better solution in a coffee than, say, sugary milk. Breves became my new favorite--I could hardly believe that this option was better than a traditional latte, because it was *so much more delicious.* I'd cracked the code. I could not, then, understand why I couldn't lose weight nearly as quickly as before, despite making dietary changes and exercising regularly. I had to call that shareholder meeting: what are our inputs, what are outputs, and where are we falling short? Well, it turns out that using half and half also meant I was dumping around 700 calories and 40g of fat into my body *every morning.* No matter how you shake that out, it was monumentally more than I had thought. So the breve was exchanged for a black coffee, and suddenly, back on track. I needed accurate information about what was happening to address and recalibrate. This example is incredibly simplistic, and yet still gets to the idea of root cause analysis. If I am experiencing deep depression, toxic relationships, and generally not getting to a place of sustained health...I need to find the breve. I need to find what is hurting how I operate within the world. This is where traditional approaches fall short.

It's not simple to identify an internal behavioral root cause. This is largely because the causes are multifaceted, just like their expression in daily life. To make it even more challenging, *all trauma events are individually experienced, understood, and managed.* Remember epigenetics? How you handle stress is tied directly to your genes, the environment, and the interplay between the two in genetic expression. This can start unbelievably early, too--before you're even breathing the air outside your mother's womb.

In researching trauma, the rabbit hole rapidly becomes a tumble a la Alice's descent into the madness of Wonderland. I became obsessed with understanding it as a result of working with individuals with chronic disease, magnified with the astounding realizations offered by Adverse Childhood Experiences (ACEs) driving concepts of trauma-related symptom sequelae, and variably applied in educational settings. Working in detox and behavioral health crisis intervention, it was obvious that a common thread was hardship in childhood--fractured parental relationships, death of a loved one, emotional abuse, sexual assault, and more. I noted in chronic disease case management that the compounding nature of illness seemed to always have a hold on clients--they lived forever in poor health, and it really, truly seemed impossible to climb out of. It was an identity. There was always a setback, a new issue, another illness. In providing basic and technical education to disadvantaged youth and young adults, I observed trauma-response in the classrooms, ranging from defiance to self-sabotage to outbursts. I also noticed something particularly interesting--incidences of ADD/ADHD and autism diagnoses were far higher in the student populations with histories of adversity. Abused and neglected kids were far more likely to struggle in settings requiring organization, self-control, meeting challenges, and cognitive flexibility. Summarily, kids with trauma did really, really shitty in traditional classrooms.

This is evident in the years of research before and after my time in education and educational administration--the field of trauma-informed care and education has exponentially grown. I started my research from the perspective of motivation within the classroom--how do we motivate students who don't want to be there, struggle with expectations, despise authority, and have no foundation of trust with any adult? It rapidly became not a question of motivation, but understanding self-regulation--how do we help students down regulate huge emotional responses, and up-regulate under performative responses when apathy and malaise take hold? Recall earlier when we briefly touched on the *nucleus accumbens*, the neurological location of the seesaw that controls inhibition and motivation: this is where trauma really, truly gets its claws in. It hijacks the highs and lows of dopaminergic response mechanisms because the inordinate amount of stress (largely summarized as cortisol) leads to a severe depletion in serotonin and dopamine. Excessive cortisol even damages the receptor sites of those neurotransmitters--making it harder and harder to balance internal responses to external stressors.

The overproduction of cortisol in pregnant women, for example, has been observed to diminish telomere length in their offspring. Telomeres are protective caps located at the ends of chromosomes, which are structures in our cells that contain genetic information (DNA). They consist of repetitive sequences of non-coding DNA (in humans, the sequence is typically TTAGGG repeated thousands of times) and are bound by specialized proteins. They offer chromosomal protection, as the capping of the ends of chromosomes prevents DNA damage, which leads to chromosomal instability or even fusion with other chromosomes. They also help prevent loss of genetic information--when cells divide, DNA replication can't copy the very ends of chromosomes, the telomeres act as a sort of buffer zone--the loss of vital genetic information is limited, and really only the repetitive sequences are lost. They also do a host of other things, including regulating cell division. When cells divide, telomeres shorten with each division--when they are too short, the cell stops dividing or undergoes a programmed cell death, so there isn't damage to the DNA.

Aging impacts telomere length, as cells divide over time, and shortened telomeres are associated with age-related disease. Things like stress, smoking, obesity, and lack of exercise contribute to the shortening of telomeres. The interplay of telomere length and disease is an area of burgeoning interest--from cancer to dementia, research increasingly reveals that telomere health is a factor in overall health. Now recall that severely stressed pregnant women are producing way more cortisol than normal, which is passing on to her unborn child. Elevated cortisol levels due to trauma or chronic stress can negatively impact brain development in children, particularly in areas like the amygdala and hippocampus, which are crucial for emotion regulation and memory processing. Cortisol-related disruptions in the prefrontal cortex can lead to reduced cognitive flexibility and impaired emotional regulation, and these maladapted stress responses can exacerbate behaviors like hypervigilance, compulsions, or emotional dysregulation in individuals who experience early trauma.

Persistent high cortisol levels can induce dendritic atrophy and reduce synaptic connectivity, weakening the neural architecture that supports coping and adaptation. These effects are more pronounced in individuals who experience interpersonal trauma (e.g., abuse or neglect) as opposed to non-interpersonal stressors (e.g., natural disasters). While non-interpersonal stress and trauma certainly produces problematic stress responses, those responses differ and have varied applicability later in life. It's important to note, however, that chronic and treatment resistant disease is typically an outcome of complex trauma--more than one hardship, more than one concept, more than one maladaptation.

Let's humanize this a little. Imagine you have a pregnant woman, and her husband suddenly dies of a stroke. The financial and emotional stress of scarcity, grief, and loss compound as time marches on, dumping cortisol into mom. When the baby is born, they are hyper reactive--they cry at everything, they're hard to console, and it becomes readily apparent that mom is struggling. She can't produce enough breast milk, and this is largely due to stress and having to work odd hours to make ends meet while the baby is shuffled between sitters. Baby's attachment is now impacted--the difficulty soothing them, the struggle to breastfeed, and the inconsistent time with mom--all of this creates higher levels of stress for an already stressed baby and mom. If you could look inside the baby at its telomeres, you'd see they are shorter than other babies, and those other babies seem to self-soothe more effectively, display less anxious attachment, and are more open to comfort and co-regulation of emotions with their caregivers. You can infer the cyclical nature of stress in this scenario--the relationships mom has with other partners, the financial strain and absence from home to generate income, the potential substance misuse for mom to help with the grief and loneliness felt with the loss of her husband. This situation, from the prenatal environment diminishing cellular health in the baby, creates a deficit in which health and wellbeing is further and further from reach. Life isn't easy, and it doesn't get easier with a severely limited ability to manage stress.

Mismanaged or poorly managed stress compounds the stress events. Behavioral patterns that result from adapting to stress become solidified with maturation, creating future issues in managing hardship and adversity.

Let's humanize this again, but differently. Imagine that a family experiences divorce. This is incredibly stressful for both the parents *and* children, as the events and strain leading to divorce hardly begin when the papers are filed--they are typically a long time coming and rife with conflict. Even in normative divorces without high conflict, children experience a fracturing of identity, as the parental unit significantly informs the identity of the child (for better or worse). Couple this with the financial stress on both parents, and the potential pitfalls of new dating partners and increased parental use of drugs or alcohol (very typical in early stages of divorce). It would be reasonable, then, that the kids now dividing time between two homes and experiencing parental divorce fall behind in school. Now here is an interesting indicator: children that are behind by a *singular grade level* in reading by third grade are 16% less likely to graduate on time. If that child is living in poverty, this jumps to a 22% reduction in graduating from high school *at all.* Project out how well kids in poverty that have no high school diploma do in life overall, and you see how aggressive the cycle of disadvantage and hardship becomes. The impacts of childhood adversity as simple and common as divorce in early childhood can last for generations, and much like gaining versus losing weight, its way harder to be healthy when you start at a marked disadvantage.

It is not, however, impossible.

The neurological impacts of trauma are also the code to the healing from it. When we look at synaptic degradation, dendritic degeneration, and cellular breakdown, we must also hold in our hand synaptogenesis, neural plasticity, and cellular maintenance. For the baby born in the wake of their father's death, disrupted attachment and maternal stress do not have to be the indicators for all of life. In fact, early childhood and adolescence are two periods of critical plasticity--two distinct developmental epochs that offer opportunity to develop new neural pathways and mitigate the impacts of trauma response.

For teens with histories of hardship, experiencing consistent, empathetic caregiving provides significant improvements in mitigating both internal and externalizing behaviors. For children struggling with emotional regulation, living with a healthy adult or older sibling that helps co-regulate big feelings and encourages critical thought promotes improved self-esteem, self-efficacy, and management of emotions. A strong, empathetic teacher that supports reading improvement and achievement can fundamentally alter the educational trajectory of students well behind grade level. Recall again the child who holds the same thought shared by all children: I want to be loved.

The difference a strong caregiver makes in the life of a child or adolescent cannot be underscored enough. This is precisely why equitable parenting is so crucial to mitigating the trauma of separation and divorce for children. State interference through unfair child support determinations, alimony requirements, and poorly conceived "primary parent" designation is almost always biased against fathers, but is largely a racket in and of itself. Criminal assignment of child support arrears while civil designation is reserved for violations of parenting time is just one of many examples in which the family unit is decimated in a courtroom and children are left holding the trauma in their brains, bodies, and development. The vast majority of adult children of divorce that did not have liberal and equitable access to their fathers blamed *their mothers* for withholding that parenting time. Across the board, children with unequal time longed for more access and relationship with the minimally included parent. For children growing up without fathers in the home, the tools and identity development associated with paternal child rearing are withheld. For boys, correlation between reduced or removed fathering and oppositional defiance, uncontrolled aggression, anger, violence, poor self-esteem, academic failure, and substance abuse are readily observed. For girls, correlation between reduced or removed fathering includes early pregnancy and risky sexual behaviors, substance abuse and dependence, obesity, poor self-esteem, and mental illness. These correlations beg the question, though: if we can take health away by removing consistent structure and caregiving, can we put health back *with it?*

In many cases, yes.

It is the beauty of plasticity that the harms absorbed might be reformed and adjusted to healthier functioning. It's why therapy can work. It's why medications can help. It's why we love stories of overcoming hardship, developing resilience, showing grit, becoming something greater than your circumstance.

It requires, though, the time and ability to redevelop aspects of trauma and stress impact to better regulate processing hardship and adversity later in life. For individuals with chronic, complex trauma, the opportunity to reinvent the neurodevelopment that was hijacked by toxic stress becomes less and less available. If you have a cold, your body needs to process it, find opportunities to maximize health, and fight it off better next time. If you have cancer, you cannot simply process it, let it run its course, or have hot tea and lemon to combat it. You won't fight it better next time.

Complex trauma leading to chronic, treatment resistant disease is the behavioral health equivalent of cancer. Individualized, opaque in origination, identifiable in harm, and a nightmare to get rid of. In fact, aggressive cancer often requires a scorched-earth approach: decimate everything. Start over.

Reset.

5

SOME KIDS TURN OUT ALRIGHT

This book is a labor of love, fascination, and advocacy--I truly believe that addressing deep trauma experiences in individuals can help chip away at the collective poor health of our nation and build back an identity of fortitude and innovation. A lovely separation enjoyed in academic and fictional writing is one between the personal and the professorial...yet the labor of love and advocacy can create improved impact through personal context. I'm going to take this singular chapter to provide personal context, in hopes that honest communication about my own experience may provide an additional bookend to the research informing the entirety of this work. At minimum, I hope it helps shed light on the perceptual nature of stress, and the idiosyncratic ways we manage and develop despite and because of it.

The pacific northwest of Washington State is immeasurably beautiful, and growing up at the base of the Cascade Mountain Range was idyllic and a child's dream. Rivers, room to run, endless opportunities to be muddy and wild were the features I remember most from my childhood. Other features were less idyllic.

My biological father had the misfortune to be significantly abused and neglected as a child. The very little I know about him can be boiled down to this: he had blonde hair, blue eyes, stood around six feet tall, worked for Boeing, was a church leader in our congregation, and had an unpredictable temper. He did not, to my knowledge, drink alcohol or consume drugs. He fathered my three older siblings. He married my mother when she was around nineteen. My only other memories include his isolation of children from their peers and parents, his careful admonition to stay silent, and his ongoing and chronic need to molest and rape children from church, our rural neighborhood, my siblings, and me. I had the benefit of being very young in experiencing this, with only vague recollections of being guided upstairs to the attic room with its singular naked light bulb swaying as my blue corduroy overalls were unfastened. When my childhood best friend, our neighbor, was brave enough to say something to her parents, the dam broke. Abruptly, he was gone. The police gathered information, we were interviewed by counselors, and no more friends came over. I think about this now from my mother's perspective, and for all the judgements I have previously assigned on our relationship, it is abundantly clear to me that she carried the unbearable weight of guilt in a way few could imagine.

She didn't know. She didn't know, because she wasn't raised horrifically. She was raised well, and cared for, and loved. I don't think it ever crossed her mind that she'd married a monster, and I am certain, though we have never discussed it, that only in reflection does she see the flags and indicators. In real time, the stress of a volatile and manipulative husband, going to college to be a teacher, raising four children, and living in intense isolation and poverty likely overwhelmed any suspicions that things were not quite right. The unbelievable decade and a half in which he hid his evil habits from her must have been immeasurably heavy, all at once, settling on her shoulders and demanding recompense. The few times this has ever been talked about stand out--primarily, when she'd done her very Christian duty, and went to where he was held in custody, and asked for a reason, an explanation, anything. Divorce is unheard of in the fundamental religious mind, and she was searching for anything to make this reasonable, or to determine it was a huge misunderstanding. His response was, "Oh, it isn't that bad."

I consider my siblings often in this scenario. My brothers and sister endured far more than myself, as our biological father was arrested when I was four; my older sister is around four years my senior and my brothers around 6 and 11 years older. My stepfather came on the scene very shortly after--I believe a horrific divorce on both his and my mother's sides was a strange opportunity to seek safety and solace in one another. My stepfather arrived to buy a truck from us--he needed it for his landscaping business and my mom needed the cash. His gigantic, fluffy German Shepard mix, Gus, jumped out of his car and refused to get back in. There were kids, a yard to run in, and it was as though he knew we needed a floofball in our lives more than ever.

There is an odd tone to my childhood memories, almost a photo-enhanced color quality to the grass and trees, garden and river, sky and clouds. Eventually, the roof got finished, electrical was completed, the walls and floors inside were closer to finished. Our little rough-hewn box-cabin expanded, as did our family--my little sister was born five years later--and my older siblings moved out, and moved on. I floated through those years until the abysmal adolescent phase of internal discomfort, external insecurities, and the typical developmental milestones of angst and reinvention. Consistent disagreements about faith, religion, and the world led to seeking solace in a friend group experiencing a shared dysfunction in their family lives. One of those friends would later become my first husband. But that's another story.

At sixteen, around springtime, I'd been permitted to spend the night at my friend's house--let's call her Erin. She and her sister were largely unsupervised--no real relationship with their dad, and their mom would have us girls sign her sponsorship forms as she navigated addiction and disease resulting from it. Their mom was not particularly thrilled with us girls, for reasons I can't recall, and we were booted from the house quite suddenly and unexpectedly. In my mind, stating that Erin's mom wasn't stable and wanted us to leave meant that I would be disallowed from ever returning--my parents let me out hesitantly and rarely. I opted for simply staying at our other friend's house--let's call her Louise. Somehow my mom found out--she'd called Erin's house, perhaps, and then immediately called Louise's. The culmination of teenage angst, dishonest communication, and general difficulty getting along with my parent's rules led to a heart-thudding button push on the phone, effectively hanging up on my mother.

"I can't live in that house anymore. I can't. We should run away," I said.

Louise and Erin, along with Erin's little sister, got bright -eyed and enthusiastic. It was settled--we'd gather what we had, we'd load the car, we'd drive anywhere, we'd get jobs, we'd be fine. It took less than four hours for Louise's mom to find out, and convince us to just come back to her house and have pizza. The plans to be alone and independent disintegrated into a general sense of "maybe when we're older", with the exception of me. I couldn't go back.

Whether my parents remember this as I do, which I doubt, there is one theme I think is held as true by both myself and them: I wasn't easy to live with, and neither were they. The ability to navigate strict rules and expectations, deeply-held religious beliefs, remain organized and responsible, remember critical tasks, events, and duties, and generally pay attention were things I struggled mightily with. Family doctor visits for medication, counseling from Christian counselors, and a lack of friendships in our isolated, rural area compounded my belief that I was not going to be understood, and that if I was indeed crazy, I'd rather be crazy on my own. On a Sunday when I knew my parents and little sister would be at church, I drove nervously to the house to collect what things I had, and saw they had been placed in garbage bags and left on the porch. The door was, for the first time in its history, locked. The locked door was significantly more hurtful to me than the garbage bags--I knew they didn't want me at home. But the idea that I would rob them, or hurt their home, or otherwise be harmful...it struck me that perhaps I was right. I was not just misunderstood--I was deeply unknown to my family, and unwanted. Fortunately, my friend Louise and her wonderful mother welcomed me to their home, halting the revolving couch surfing and all that accompanies it.

Louise's mom (let's call her Libby) was quirky and bohemian, and permitted far more autonomy than I'd ever experienced in my hyperregulated home. When I look back, I identify patterns that revealed the transactional nature of love that I believed was true. My biological father laid a foundation of extractive love: love is what you let others take from you, however they want. I was always messing up at my parent's house--I couldn't remember things, couldn't stop being impulsive, and struggled to answer things truthfully and clearly. I couldn't stop asking questions, all the time, and it was these things I believed I did so poorly that I was not worthy of being kept in the home. With Libby and Louise I tried so hard to be worthy--I'd do dishes, I'd take out the recycling, but still, organization, impulse, and anxiety took hold. I could, however, work. I was really, really good at that. Libby was the Director of a respite house nearby, and coached me on how to interview well--the day I turned 18 I was hired. By this point I'd long completed the ridiculous packets from the alternative school and graduated, and fulltime work was the next logical step--no more cart pushing and babysitting, I was eighteen, and in my mind, this was a new adventure.

A crisis center is one of the most interesting places to find gratitude. It would have been normative for me to look inward and think negative, nasty things about my situation, but it isn't my nature and I truly think working there was a critical path that enhanced internal fortitude. This crisis respite was a six-bed house, unmarked, double-staffed on 12-24 hour shifts, offering basic detox and mental health stabilization services for 1-5 days. The regulations and policies needed now for these facilities are far different than when I cut my teeth in behavioral health--making for some fascinating, if not high-risk interactions.

I walked into the house through the backdoor, which led straight through the kitchen to the rest of the house. In the kitchen, standing awkwardly near the oven, was a fairly tall man in his thirties. He looked quite average in most ways, save for the fact that his pants were down and he was doing his best to quietly crank up the oven with his penis awkwardly shoved in the side of the door. We locked eyes, and with a deflated sigh, he pulled up his pants and trudged into the living room. My shift partner, frazzled and pink-faced, came bustling through carrying a number of scissors.

"Are you the new person?" She asked without preamble.

"Yes, I'm Melina," I said as confidently as possible.

"That's Jeff. Don't let him have anything sharp. He's always trying to cut off his dick."

I nodded. "Ok, um, I think he was also trying to burn it off in the oven," I said. She rolled her eyes and trotted into the kitchen, dumping the scissors into a locked drawer and turning off the oven.

"Welcome to the respite house," she said. And that was that. Jeff was one of many severely, chronically troubled clients that came in and out of the respite house for stabilization, step down from hospitalization, brief detox services, and transitional placement. I had been raised so strangely--first, intense violations in formative childhood; this was followed with strict religious belief, and then sudden expulsion from that sheltered and isolated world into a reality in which people tried to burn off their penises. A world where young women my age would cut their own throats on a smoke break on the porch and wait for you to find them. Where old men drink bleach to wash out the bugs in their stomachs, or young men fight off Scottish warriors attacking them from the chests of strangers on the street. A world where people were so broken, so ill, so hurt, that it was impossible not to ask how does this happen?

I worked two jobs for much of my young adulthood. From eighteen onward, I would max out at 80-85 hour workweeks, scheduling appointments at a radiology clinic Monday through Friday, and working 24 and 13 hour shifts over the weekend at the crisis respite house. Once a week I simply gave up a night of rest. I had the itch to travel, and eventually bought a one-way ticket to Dublin after spinning a globe and putting my finger down on Ireland. Spending a few years in Ireland blew my mind open to the world: people had fun in the world. This was a revelation! I had never really considered fun to be anything worth pursuing, and now I was nineteen, at a hostel filled with European young people like me. Accents, food, culture, histories, exploration...it was all intoxicating. And yet, these same tricky patterns kept emerging: I would seek out exploitive relationships, conduct myself impulsively, forget critical information, struggle to remain organized, and generally always felt I was looking through a window at everyone else, wondering what it would feel like to be on the other side. It was the Looking Glass, but I was in Wonderland yearning for the traditional English countryside of predictable expectations and acceptance. I was, of course, the child wanting to be loved. I didn't know how, and I didn't know what it looked like.

I came home to Washington with an idea of returning to school. I'd watched young people my age walking to Trinity College in Dublin, and all I could think about was how much I missed learning, and missed being in the classroom. I knew I needed some sort of education, so I enrolled in a community college with the trajectory of fine arts--I loved to paint, draw, play music, and sing...maybe that would be a good life for me. It took a semester of being among "artists" that didn't wear anything on their feet in class and had millions of thread-thin scarves tied fancily, and I simply couldn't buy in. I tried graphic arts, but the idea of someone making you change your work to sell something sat poorly with me, too. I was back working for the crisis center, now in a newer, bigger facility, and that's when I encountered my very first mentor. The first person I'd met that wanted to see me do well, was willing to carefully guide and support me, and didn't want anything else. He just saw potential, and goodness, and told me I was capable of running the center someday. This, along with my coworkers, became the first time I really felt I belonged, and that I was wanted. I was good at something, and that something was creating, operating, and evaluating programs that helped people.

I enrolled in a Human Services program. A professional helping program--I'd get a degree in helping. From a child unable to do the right thing ever to a professional helper seemed too good to be real, but the undergraduate program was precisely what I needed, when I needed it. I had reconnected with my old friends, Robert and Louise, who detested each other now. Robert and I had always had a connection through reading and clever smartass dialogue, and eventually, we moved in together. He was always challenging--whether being hurtful in his cleverness or unreliable in his friendship, he was a vacuum of sorts for a person accustomed to always pleasing and helping others. Not knowing how to be loved is an excellent way to partner with people that don't know how to love others. I saw his family, though, and believed maturation and finding success would take care of the other issues--really not considering the fact that he was uninterested in maturing or being successful in any way. I was immediately the breadwinner, the student, the caretaker; eventually, I was the adult, the mother, the graduate student, the professional, the university educator; at home, the punching bag. For each success, outward, social celebration and pride in being a "house husband" translated to harsh criticisms, slaps, arms twists, and forced physical submittals for his delight. It didn't matter if I was pregnant, exhausted, or firmly uninterested--if he wanted sex, it was required. If he wanted weed, it outweighed rent. If he wanted my hair red, it was red. If he wanted me to avoid any workgroups that had males, I had to. No social media. No meeting off campus for school or work. No lengthy stays at Louise's, and absolutely no meaningful contact with my family, particularly my sister. When my daughter was two, it was clear she was computing his treatment of me, his withholding of affection from her, and his disinterest in ever working. This was being presented to her little mind as normal behavior for a partner with my acceptance of it. When he shot out the glass slider while high and cleaning a gun with his friend, it was past time to go. We were not safe.

He didn't like me to have the car when I worked. At one point I was walking to my teaching job at the university, eight months pregnant, up and down the hills leading to campus. I vomited the entirety of both pregnancies, and when I sat in my wonderful senior colleague's office, sweating, nauseous, and exhausted, the dam broke. I burst into tears.

"I feel horrible. I feel horrible because I think how much easier it would be if he just got hit by a bus, or had a heart attack or something.

Then I wouldn't have to do this anymore. I'm an awful person," I wept. Having been through a wicked divorce herself, she looked at me through her big lens glasses, her short grey hair wispy and her tiny frame leaning toward me with absolute compassion.

"Oh honey, you're not a bad person. We'd all rather be widows than divorced."

I laughed. One of those dry, bitter laughs that comes genuinely and necessarily when faced with impossible circumstances and a terribly true joke. It helped. It helped to know that I wasn't the only person trapped through the same stupid choices I'd been making my whole life, just trying to sort out how to be loved. When I got the nerve to leave, I took my babies to Louise's house, now married with three kids of her own, and never looked back. Robert remained uninterested in being anything akin to a meaningful father, and after failed attempts at supervised visits, he simply vanished. No response to attorneys, communication with his children, cooperation with the investigator, nothing. And it was absolutely the most precious gift he had ever given me: my children, without him.

So here we find this rather accelerated path of discovery leading to a sense that, as my dear sister puts it in relation to men, that my "picker is broken". My ability to discern what is healthy and what is attractive, what is love and what is control, what is generative and what is debilitating...was significantly questionable. Again, my sister's influence was monumental. She had adopted, and her son was diagnosed with ADHD. He'd recently been upped on his medication, and the remaining low dose was still in her house. "Maybe you should try one and see what happens," she said. So I did. And the world went silent.

It was the first time I noticed that my pressured speech, my impulsive actions, my inability to stop jumping from task to task, my hyper focus on some things and absolute avoidance of others...all these things were not helping me, and they weren't typical. I asked my mother when I had just turned 30, graduated with my Master's, and was approaching a new, much more challenging job: "do you think I have ADHD?"

She chuckled. "Do you think you don't?"

An adult ADD/ADHD evaluation resulted in a revelation: I'm not mentally ill. I have issues with executive functioning. It's not anxiety, depression, bipolar, or psychosis--all diagnoses threatened, assigned, and/or insisted upon by others. I simply could not prioritize the chaos in my head. When I was prescribed medication, it provided an intense clarity that was uncomfortable, quiet, and profoundly helpful. Earlier, I said I floated through childhood; adolescence was dissociative; early adulthood felt like looking through a window at the word. At thirty, I felt like I woke up. Everything until then had been a dream. It was time to set some goals, I thought. My goals were simple: make $100,000 a year. Buy a house. Complete my PhD before I'm forty.

My children and I were alone by choice for six years. We were poor, and then I got a much better job that took me away more but paid enough for me to buy a house. Two goals, accomplished. I was beginning my PhD. We were happy. I was lonely, I calculated, approximately every 90 days or so, and that passed within a week. It was manageable. I did not, absolutely did not, trust my ability to understand what a good person who loved well would look like. So I loved my kids, spent time with my sister and my friends, worked like a dog, and built up my life based upon staying safe from my own disastrous romantic decisions.

It was during my work as Dean at a basic education and technical training school that I noted those patterns of executive functioning and other forms of neurodivergence and their connection to trauma. I had never considered myself traumatized--it was not an identity or issue I struggled with, and my approach to mitigating the residual issue of poor picking was to simply not pick. Yet I saw myself in these young adults learning and struggling at this school--they lived there for up to three years, and the classroom issues, social hardships, impulsivity, self-sabotage, and dysregulatory aspects to their experience on campus was a shock to me. It was me. It created a drive to understand why hardship is so connected to dysregulation. I also needed to know why I became Dean of a school, and the people I had worked with in the respite center committed suicide, struggled with addiction, and carried mental illness with them as an oppressive cloud consuming their daily lives. What, if anything, was the difference?

I looked at those students, and I saw the fork in the road. One led to death and disease, and one led to health and happiness. The divergence was the unknown element. Why are some kids alright, and some aren't? Why did some people experience ostensibly less hardship, and turn out worse? And what in the world did ADHD have to do with it?

The thread between trauma and ADHD (executive functioning) is not unknown, but it is underexplored. As we previously covered, the connection between neurological health, regulation of emotions, and cognitive flexibility is one impacted by the epigenetic nature of development--whatever ingredients you have will bake into different versions of expression based upon environmental factors. Evidence of positive impact consistent, loving caregiving has on children and adolescents struggling with trauma-induced behavioral issues is abundant...yet when I reflect on my own upbringing, there simply wasn't warm fuzziness. There was, however, a common theme through all of this: self-regulation.

My parents were creatures of routine, even severe adherence to routine. Dinner was at six. Jesus is Lord. School is a priority. Chores will be completed in the entirety and to precise specifications. You will be clothed at all times, modest in appearance, eat everything on your plate, and so on. There was no wavering from this--it was an incredibly powerful routine that did something I could not see at the time, and only recently understood. It was an externally forced self-regulation environment that posed hardship and salvation simultaneously.

The statistics on outcomes for children of sexual abuse and poverty are abysmal. Generally speaking, victims of child sexual abuse are four times more likely to abuse drugs and alcohol, experience PTSD, and are around three times more likely to have major depression. Adults who have experienced childhood sex abuse are less likely to complete high school, more likely to face financial hardship, unemployment, and rely on welfare. Children, youth, and adults with this history are also nine times more likely to enter the criminal justice system, and across the board, chronic disease and disability are exponentially more likely to occur. Although not a pristine statistic, it's estimated that around 35% of male sex abuse perpetrators were similarly sexually abused in childhood--and that doesn't include those that don't report, experience different kinds of abuse and neglect, or do not believe the abuse experienced qualifies as criminal or abusive in and of itself. This is what is so remarkable--my brothers should have been terrible men. My sister and I should, statistically, be repeated victims. Yet we present, in a strange longitudinal and informal case study, as remarkably healthy people. My oldest brother has been married over thirty years, with two wonderful adult children. He, his wife, and my niece and nephew are close to one another, well educated, intelligent people with normative issues typical of all families. My other brother was once divorced, for good reason, and strives daily to be the best father and husband possible for his new wife and adopted child, in addition to the children from his first marriage. He works harder than most humans I know, and has built a life for himself twice following the disastrous divorce in which his ex-wife revealed patterns of control and dishonesty that could have destroyed him--but didn't. He overcame it. My sister owns a Montessori preschool, cares diligently for her chronically physically ill husband, has two wildly healthy and wonderful adopted children, and lives with a touch of OCD that we like to call "the best possible kind", because it's light switch and doorknob checks, but it's also careful business practice and consistent cleaning schedules that provide the critical structure needed for her family. I'm a second-time-married, blended family mom of five, with a successful career and looking at a bright future...none of us has substance abuse issues, questionable parenting, criminal histories, or failed education. All of us graduated. All of us are living, if not thriving, in lives of meaning. We are not perfect, and we are not broken. We're alright.

Contrast this with the statistics and outcomes that were far more likely. The reasons behind generational trauma and cycles of perpetuated abuse are rooted in the epigenetics of trauma impacts. The stressors experienced by victims of repeated sexual abuse are like a terrible gem that turns slowly and constantly, revealing different violations and related adaptations. Ingredients to this abuse are simultaneously individual *and* generalizable. Intrusion without permission, adult self- pleasure from prepubescent children, and the physically harmful aspects of sexual touch on nonsexual, vulnerable bodies is one side to the gem. Another is the psychological manipulation requiring secrecy, of hiding abusive and profoundly inappropriate adult behavior while maintaining the deception that the activities are acceptable. A third side to the terrible gem is the identity malformation that accompanies physical and psychological sexual abuse--the altering of personal value, worthiness, safety, cleanliness, and position within the family, and within the world. The outcomes for many children of sustained sexual abuse range from intervention indicators to lifelong related conditions. Many children, prior to intervention, display signs of sexual abuse through frequent, developmentally inappropriate incontinence (bed wetting); refusal to remove certain items of clothing (such as wearing a heavy coat on hot summer days); inappropriate peer touching during play; eating irregularities; poor hygiene; outbursts at school. The list goes on. You can easily infer the reasoning behind all of these behaviors when looking back--children, despite their worldview being immature and underdeveloped, have physiological responses to gross violation. Soiling oneself, poor hygiene, and heavy dress are tiny subconscious prevention measures to physical intrusions--become dirty, become unappealing, become harder to access. Mimicking of touch, depending on the age of the child, is often an exploration if the behaviors that feel different and uncomfortable are actually normative. As abused children develop, they recognize whether it is okay or not, and future behaviors range from avoidance all together toward maladapted psychopathology around all sexual behaviors. The outbursts and eating irregularities reflect the inability for the child to control what happens to them at home, putting them closer to the edge emotionally and more likely to try to control small things, like how much they eat.

Outcomes for children differ depending upon the level of abuse, chronicity of it, whether intervention occurred, and whether adults are held to account for their actions. Horrifically, many children who are abused by a parent in this manner know only that this is their parent--they may lie, become upset, or insist nothing was wrong to protect their family member. It's also why they don't report it. The outcome of protecting the perpetrator lays groundwork for continued protection and rationalization of abusive behaviors in others, informed by a sense that your personal autonomy is not valuable. This is a hallmark of abusers: remove the identity, sense of control, value, and freedoms from someone, and they become your toy. Your plaything. Co-dependence, eating disorders, drug and alcohol abuse, major depression, anxiety, compulsive disorders, and PTSD are all attendees at the dark party that is living without your soul intact. And that's what sexual abuse does to children--it tears apart their souls, their sense of self, and their value. That person becomes someone who allows themselves to be harmed, or harms others before that can happen.

It is still shocking to me that all of my siblings and I sit well within the bell curve of healthy humanity--imperfect but intact. The only thing I have been able to discover is that we have all had to develop self-regulation, and we called on our upbringing of externally applied regulation (church, school, chores) to map structure onto our current lives eschewing trauma-related outcomes. I reflect now on the critical period of plasticity in adolescence that we were so very blessed and protected to have, if even for a little bit. The intensity of routine, the refusal to permit chaos, all of it seemed oppressive and awful as a teen--and yet, here all of us sit. Healthy. Resilient. Our trauma was not resistant to treatment, because we had the opportunity to recalibrate, to some extent, the neurocircuitry during the last moments as the mud dried around us on our developmental path.

This is the point in which we have wandered away from the idea of a reset, but we are not lost. Fear not. This is all for something: the concept of structure and routine outwardly applied reconfiguring the internal schema of the individual engaged within it. I resisted it mightily as a kid--I did not feel I needed it, I hated being told what to do, and found the idea of religiosity and adherence to principles passed down rather than discovered to be ludicrous. I missed the love I was wanting from my parents, because it looked like despotism and judgement; it was, in reality, boundaries to keep me alive and functional. A cage is a prison if you long to be free, and a safe haven if you are vulnerable. The tricky part is understanding which one you are, or navigating the issue I faced: I longed to be free, and I was vulnerable. My parents did their best to keep me safe, even from myself. When I think about love and all its manifestations, I realize that the strength and depth of investment to keep putting a naive bird back in her cage was how my parents loved me. They wanted to press upon me the true nature of living well, even if imperfect in achieving it themselves. It wasn't boundaries to create a box: it was boundaries to create a path.

This is not available to many people. The intensity of structure provided by my parents was temporary, and I didn't revisit such structure and boundaries again until I was a parent myself, and even then, it wasn't until I managed my ADHD that routine became less of an ugly word and more of a salvation mechanism. I make a list every morning: what needs to be done personally, professionally, creatively. Meals are planned, because when I randomly grocery shop, I forget what is at home. More than once, I have put crackers away only to realize I've purchased the same box of crackers four times already, and because they're in a closed cupboard, I forget they exist. I keep a huge white board in the kitchen, with a calendar and reminders, a dry erase grocery list, and all the kids have chore charts for morning, afternoon, and evening duties. There are clocks in every room. I go to the gym at lunch every day. Dinner is at 6pm. It's not as tight as I'd like, but we also have movement with my three bonus kiddos flipping weeks between our home and their mother's. Flexibility is needed. Routine is required. My daughter is a nearly carbon-copy of me: she needs this as much as I do.

If this was not provided to me in childhood, I could not have trained myself and my family in regulation like this, either. Without medication, I could create all the structure and conceptualization of routine and would navigate it with around 40% efficiency and 1000% stress over the remaining 60% that I can't grasp hold of.

Children who do not get this kind of structure, combined with compassion, face compounding symptom sequelae for traumatic events and the impact on their ability to self-regulate. This is where trauma-related disease and disorder must be intervened within that critical plasticity period: early childhood or adolescence. This is where patterns of regulation and response to environment have more squish, more opportunity for neurogenesis to address trauma-impacted processes of the brain. When this period ends...the ruts are dried. The path is set. And nothing short of miraculous intervention or lifelong effort will augment outcomes for victims of trauma-related, chronic, treatment-resistant disease.

This is the extent of personal history that I'll share, with tremendous gratitude for your patience. We can move on to concepts of change, cycles of destruction, and the critical importance of cognitive flexibility and emotional regulation in maintaining consistent behavior long enough to enact change.

6

I'M SCREWED AND ALL I HAVE IS A HAMMER

One of the biggest frustrations in treating chronic conditions is the seemingly impossible aspects of constant relapse and recidivism to destructive habits and symptom expression. Particularly frustrating is watching a loved one wither away into addiction, with perhaps the most common and heart-wrenching circumstances a result of opioid addiction. Generally speaking, there are four waves of the opioid epidemic in the United States. The first wave, largely in the 1990's, is characterized by the over prescription and overutilization of prescription opioids, introducing severe addiction as a result of pharmaceutical pandering and manipulation of prescribers, patients, and society in general. Hailed as miraculous pain removers and life-restorers, companies like Purdue Pharma reached their gnarled hands of greed out from their previously wildly profitable "mom's helper" Valium and into the world of chronic pain management. There was not only an outright disregard for the lives harmed with, primarily, OxyContin, there was a concerted effort to lay blame for addiction on "the addict". The second opioid wave in the 2010's is a reaction to the rampant addiction and overprovision of prescription opioids. Summarily, prescribers got their hands slapped and stopped prescribing, with minimal support for the now addicted and hurting patients present during the transition back to fewer prescribed opioids. This resulted, largely, in use of heroin and other illegal opioids to combat original chronic issues like pain, and new issues of severe addiction. The third wave, around 2013, saw fentanyl synthetically manufactured and put into illicit drugs, leading to increases in overdose deaths across the nation. The fourth wave, which we are currently in, is largely issues of fentanyl in polysubstance abuse contexts, including but not limited to opioids, methamphetamines, and other synthetic drugs.

The victims in these waves are consistently the people struggling with addiction, their families and friends observing their destruction, and the communities in which they reside shouldering the burden of care provision for unending cycles of relapse. In 2020, opioid addiction cost the US economy around a whopping $1.5 trillion. Consider the overdoses, deaths, treatments, job losses, physical deterioration and infections, family fractures and incarceration, and this number begins to knock on the door of what the fuck. If we could reinvest even half that number in public education, infrastructure, improved wages, and reduced housing costs, many of the issues preceding addiction (poverty, behavioral health support, improved access to medical care, adequate education and housing) could be reduced or eliminated. Yet we are limited to advocacy and small governmental coalitions begging for settlement dollars from big pharma to pour water into a sieve of trauma-influenced addiction cycles. It's the analogy of being on a bridge, pulling people out of the water, and hoping funding lasts next year for more ropes and buoys to continue saving them...but no one is walking upstream to find out why people are falling in the river in the first place. Addressing symptoms is not addressing disease: treating addiction epidemiologically means we address trauma.

Patterns of destruction are present in most people, with varying degrees of severity. It can be as minimally harmful as eating late at night because it feels so good to just sit and binge watch Netflix with chips. It can be maximally harmful when it's addiction to drugs or alcohol, or engaging in chronic cycles of abusive partners. For people with adequate social support, a decent financial outlook, and a desire to rid themselves of addiction, traditional approaches to self-reflection, safe detox, community engagement, addiction counseling, and a change of lifestyle, recovery is possible. An issue faced by many, however, is that unaddressed and unmitigated trauma can create lifelong patterns of destruction, in which addiction is a symptom of residual processes that prevent self-regulation. When you are unable to self-regulate your emotions, cognitive states, and stress response, you are once more on the teeter-totter of dopaminergic dysfunction and reversion to survival brain states.

I draw these distinctions for a reason: trauma-influenced chronic disorders originated in the neurobiological processes that adapted to survive, and persist in contexts they are no longer needed. This becomes maladaptation, in which an adaptive response no longer helps its host, but rather, harms it through jumping in the driver's seat when it ought to stay in the back. It becomes the primary tool. If all stressors, from difficult conversations to job loss to breakups are perceived as life threatening, they are all nails ready for your crucifixion. When encountering stress, you bring out your hammer. It's the only tool you have developed, because your whole life, you've only ever seen nails.

Think about consumer culture a little bit. We are taught through advertising and identity-promoting media from a very young age that the appearance of affluence or influence (wealth or power) is the appearance of goodness--success, acceptance, independence, popularity. We are also groomed societally to seek external solutions to internal problems: take a pill, have surgery, get a nice house, show off your success on social media, generate an identity based upon approval from strangers, and so forth. Even over-therapizing is an aspect of this. Eternally relying upon a therapist to help you understand normative experiences is much like overreliance on your phone's navigation rather than remembering routes to places you regularly drive to. You lose the skill when you permit its atrophy, and atrophy comes when you wait for someone or something to solve things for you. It makes sense to use your navigation when you've never been somewhere--but once you know the route, you can stop relying on it. Independent, critical thought is a feature of self-efficacy; without internal navigation of external stressors, you will constantly be seeking external solutions to the internal state you're in.

Self-efficacy is a product of self-regulation in many ways. A firm grasp on identity, or who you are in the world, is a starting point for confidence, security, and motivation. Recall that motivation resides in the seesaw land of the nucleus accumbens, with its counterpoint of inhibition. Managing motivation and inhibition is regulating desire and avoidance, pursuit and restraint, and ultimately, homeostasis. Balance. Without maintaining balance, we are vulnerable to the whims of the world: we are looking for external things to make us feel good, with the outcome feeling worse when the happy stimulation is absent.

One example of this is screen addiction. Even in healthy, happy, untraumatized homes, you can easily observe the impact of removing and reducing screens from children. The highly stimulating, short-lived, entertaining, visually engaging, constantly changing onslaught of videos and imagery is fundamentally and often intentionally addictive. The impact on dopaminergic and serotonergic processes cannot be underscored enough--those happy chemicals dumping into their brains with constant content are suddenly and abruptly halted when the screen is off. It is little wonder that personal screens have become both the opiate and the bane in parenting--we are dosing up our children and observing withdrawal symptoms. If you don't have children, or perhaps you think I'm waxing poetic on this, remove a screen from a kid in the middle of a YouTube video. See how happy and compliant they are. Or ask them to sit through an entire movie, without interrupting, narrating, or losing their minds over and over again. Ask them to engage in sustained reading, with only text, and see how much they can remain attentive to the content. Addiction to highly stimulating content is a natural human adaptation; we are tasked with hyper focusing on attention-grabbing stimuli for our survival. Yet is becomes maladaptive to continuously and aggressively seek out that stimuli and ignore normative, less engaging and yet essential functions, like going to bed because we need rest. If you've scrolled on your phone in bed and realized in the morning it was a terrible idea to exchange three hours of sleep for three hours of cats being frightened into hilarious freak-outs, you understand. You will regret poor sleep, you will not regret missing videos that you've already mostly forgotten and served no purpose outside entertainment.

It is precisely this prioritization of what should be done, what can be done, and what should be avoided or limited that maintains our balance. This is self-regulation. When we are swept up in overstimulation, we enter into that vicious cycle of chasing good feelings and avoiding bad ones. This chasing is in direct conflict with consistency, as it takes consistent prioritization of wants and needs to keep regulatory processes healthy and functional.

Let's look at another addictive substance we've touched on: highly processed foods. Since the 1980's, highly processed foods have increasingly taken on the majority role in American diets, and in keeping with the second wave of the opioid epidemic, a sharp spike in both ultra-processed foods and obesity and chronic disease can be observed. Ultra-processed foods contain a number of chemicals and additives designed specifically to make foods more appealing visually, texturally, and flavorfully. They also are added to increase the shelf life, create hyper-stimulation with dumps of happy chemicals in response to sugar, fat, and salt content, and ultimately, bring the consumer back repeatedly to access the same level of pleasure unobtainable in unprocessed and minimally processed foods. It is made to be addictive, and combined with being cheap and accessible, it has disproportionately impacted the health of our population, particularly children (who have no concept of restraint yet) and those in poverty (who have no access to affordable, healthy nutrition). Rather than address the abuse of American health and wellbeing by ultra-processed food manufacturers, we have turned our eyes to semaglutide products like Ozempic, looking once more for an external solution to our obesity rather than the imbalanced regulatory processes undergirding overweight and poor health. I want to take a moment here and note that Ozempic, like all things, is an example of poison when taken too far. For some people, the quick-win or "boost" of rapid weight loss can support healthier habits and the concept health is possible. The issue is that the long term impacts of these products are understudied, and may well present dependence upon them to maintain weight loss. This creates that unhealthy seesaw once more: loss rapidly, and keep taking it or risk putting all the weight back on. A positive change that I hope very much for is increasingly loud frustration about the additives in our foods that are harming us hormonally, metabolically, and behaviorally. Prohibiting harmful chemicals to increase food manufacturer profits is a step in the right direction--yet it doesn't resolve the underlying issues of addiction, compulsion, and consumerism.

For individuals with trauma-induced behaviors, change becomes all but impossible as the circuitry for sustained regulation and adherence to protocols is absent, or misfiring. If you have only ever had a hammer, recall, everything is a nail. Navigating stress and the reality of life's hardships is difficult for all people; for those with maladapted neurobiology, it is impossible to do so in any sustainable fashion. Normative developmental contexts in managing stress and emotions provide a variety of tools at the adult's disposal. Children and youth who are consistently supported, guided, loved, and protected have years of critical development in which they are training their brains to react, respond, and regulate in relation to events, feelings, and stressors. They develop into adults with a Swiss Army Knife--they can see an upset partner as being upset and needing comfort, and they select that tool from their developmental toolkit. An adult with a history of abuse or neglect may see an upset partner as displaying behavior that will lead to harm, through abuse or abandonment. This person will pull out their threat-mitigator, the hammer, and react from a position of defense or offense--an enormously different outcome for that relationship.

Management of hardship and adversity is trainable. Much like genetic predisposition accompanied with training creates the athlete, neurobiological development with stress management experiences creates the overall individual response. The issue with maladaptive response is that for humans, we routinely just do what we know. We follow a template that has been established when our brains were learning how to navigate the world.

There are a number of positive examples of training healthy adversity and hardship response. A program that many people may not consider preventative behavioral healthcare is one that has been quietly run in rural areas since just before the early 1900's. This program, 4-H, originated in the early 20th century as an educational movement to bridge the gap between rural youth and advancements in agriculture and home economics. The idea began with "Corn Clubs" and "Tomato Canning Clubs" in the late 1800s and early 1900s, where young people could learn practical skills and demonstrate agricultural innovations.

In 1902, A.B. Graham, a school principal in Clark County, Ohio, formed the first organized youth agricultural club, which many consider the birth of 4-H. The program expanded through land-grant universities and the Cooperative Extension System, which connected agricultural education to local communities.

The name "4-H" and its four-leaf clover emblem with the Hs standing for Head, Heart, Hands, and Health, became official in 1914 with the passage of the Smith-Lever Act, which formalized federal funding for Extension services. Over time, 4-H grew beyond its agricultural roots to include leadership, citizenship, and life skills, becoming a worldwide youth development organization. I have had the pleasure of be involved with 4-H through work and my children, and became fascinated with the incredible resilience of the rural kids involved in animal projects.

For the kids raising livestock animals, particularly market animals, the process includes careful tracking of the care, finances, and presentation of livestock animals for selling, and, ultimately, butchering. These kids, as young as eight years old, are up before school and out after school caring for their animals. They groom, feed, clean, and track care and expenses in their record books, and show their animal at the annual County Fair. They are scored and placed on the quality of their animal, and the auctioned animal often brings in thousands of dollars to replenish costs incurred and provide the individual with an occasionally significant profit. Market Day is the day the animals are loaded, also called Load-out Day. These kids—remember, some as little as eight years old—take the pig, sheep, goat, cow, rabbit, or fowl that they have raised diligently to the trailer that will transport their beloved animal to their death. It goes further—there is often a carcass contest, in which their animals are graded and judged hanging in the processor for internal health and quality. This additional scrutiny of the animal displays the feeding practices and less visible health and care of the animal as it relates to marketability and desired taste and appearance.

I spoke to a lifelong 4-H mom and leader, and she made one of the most remarkable observations that has stuck with me over the years. I was providing a brief training for 4-H leaders on mental health, and she shared in discussion thoughts about resilience and managing hardship. She noted that her children managed death and dying far more effectively than their cousins, who did not do 4-H animal or livestock projects. The process of caring for, loving, and organizing the death of an animal created a template of navigating grief that was applicable to loss of family members and other loved ones. This was noticeable even for her kids—they understood and observed that they managed death, dying, and grieving more efficiently and healthfully than their cousins.

Kids, adolescents, even very young adults *need* practice in managing hardship. When traumatic experiences are not navigable because you live in them, or they are acute and severe to the point of toxic stress load, the practice is less about getting stronger through it and more about surviving them. The ideal upbringing would include some level of manageable stressors and hardships (like rearing and butchering an animal) to provide training for caring and loss later in life. It's a slow drip, an inoculation of hard things so that in adulthood, children are grownups capable of doing hard things. When hardship and stress hit less like a minor cold and more like a cancer, the lifelong consequences are deficient management of future illness events (stress). Treatment-resistance lies within the distinction between what can strengthen the immune response (resilience, self-efficacy, emotional regulation) and what creates immunodeficiency (trauma symptom sequelae, chronic disease and disorder).

This book is about all sorts of traumas and related behaviors, but it's time to focus now on the population in which trauma-induced chronic disease and disorder make traditional therapies ineffective, frustrating, and demoralizing. The distinction between trauma-related treatment-resistance and diagnoses that are navigable and recoverable with traditional methods of care is the circuitry involved, the operating system within. If the operating system (OS) is programmed or updated to survive traumatic events, it will continue operating through that design, producing outcomes through the lens of survival. For OS that are largely functional and able to process the world (untraumatized brains), new or stressful inputs are routed along programmed methods of coping. Loss, grief, heartbreak, illness, and hardship may be met with resilience, critical thought, cognitive flexibility in navigating challenges, and counseling or therapy to provide skills and maintain healthy self-worth and efficacy. For OS that are dysfunctional (traumatized brains), new or stressful inputs are routed along programmed methods of coping--hyper-emotional response, severe withdrawal and emotional paralysis, anger and aggression, or self-medication with unhealthy relationships, drugs, or alcohol. It is the intention of this work to describe how trauma impacts human functioning, and how a factory reset of your software can support the necessary OS upgrades to move from dysfunctional to functional. So let's get to it.

7

NEUROGENERATIVE TECHNOLOGIES AND AGENTS OF CHANGE

There are a variety of psychedelic substances that fall under myriad names, concepts, and historical uses. The use of psychedelics in religious, spiritual, and mystical practice is found across the planet. Use ranges widely from indigenous use in the Amazon through the brewing of ayahuasca to evidence of psychoactive substances in religious ceremonial vessels in Greece. It can be found in the use of ibogaine in Africa, as well. The Bwiti religion, for example, incorporates ibogaine as a central sacrament during initiation rites, healing ceremonies, and rituals to commune with ancestors or the spirit world. Imagery from ancient texts and artwork reveals thousands of years of wheat, mushroom, and lotus flower imagery, all associated with psychoactive substances. Wheat often had fungal ergot, a powerful lysergic acid producer that was often accidentally and also intentionally consumed for its LSD-like effects. Mushrooms range from psilocybin to amanita muscaria representations, with the familiar red and white amanita peppering forests as far north as the boreal areas of northern Russia and near the Taymyr Peninsula. Representations of mushrooms are found in Australia, Mexico and South America, North America, Europe, Asia...you name it. With enormous variety in plant-type, effects, and utilization, psychedelics have been a part of human history as commonly as art and war.

Peyote is a small, spineless psychoactive cactus known for its historical and ceremonial use among indigenous cultures. It primarily grows in the Chihuahuan Desert regions of South Texas and Northern Mexico, thriving in dry, limestone-rich environments. Its natural range is more limited, and while it remains a Schedule I controlled substance in the eyes of the US government, it is permitted for ceremonial use in tribal contexts. Peyote has long been associated with spirit journeys, healing power, ancestral connection, community-building, and rites of passage.

Ayahuasca is a traditional psychoactive brew used by indigenous cultures of the Amazon Basin for spiritual and healing purposes. The preparation of ayahuasca is a ritualistic and careful process that involves specific plants and methods passed down through generations. *Banisteriopsis caapi* is the ayahuasca vine, and the primary component of the brew. This vine contains MAO inhibitors harmine, harmaline, and tetrahydroharmine. The plants *Psychotria viridis* (Chacruna) or *Diplopterys cabrerana* (Chaliponga) contain dimethyltryptamine (DMT). This DMT is not available when consumed, as human stomachs break it down and inactivate it. It is only when brewed with the vine that ayahuasca becomes hallucinogenic. The resulting brew is a viscous, bitter-tasting liquid with powerful psychoactive properties. The experience induced by ayahuasca typically lasts 4 to 6 hours and can include visions, emotional processing, and physical purging (vomiting or diarrhea), believed to be part of the healing process. This brew consumption and ensuing intention-setting and cathartic healing are all considered sacred, and are led by shamans with generational understanding of the process and outcomes for individuals they guide.

Ketamine is considered yet another psychedelic, dissociative in nature and legal in some states for use in clinical practice. It is synthetic, and is often referred to as a "dissociative anesthetic" in medicine for its memory impacts and pain relieving properties. It is one of the few substances that has been legally used in clinical therapeutic settings in some states. Lysergic Acid Diethylamide, or LSD, is a synthetic hallucinogenic drug known for its powerful psychoactive effects, derived in the 1930's from lysergic acid from ergot--recall the wheat imagery discussed earlier. The use of LSD in countercultures in the United States as a "mind expanding" option led to the idea that the revolutionary concepts of breaking free from the military industrial complex and oppressive sociopolitical structures were dangerously motivated by "tripped out hippies". Through a series of both calculated and unexpected events, LSD became the scapegoat for all things wrong in American youth. It was removed from research, recreational, and clinical use shortly after entering society by big-G government committed to law, order, and punitive judgment of "drug users". Several psychedelics were classified as Schedule I controlled substances under the Controlled Substances Act of 1970 in the United States.

It's critically important to remember that Schedule I drugs are defined as substances with high potential for abuse, no currently accepted medical use in treatment in the U.S., and a lack of accepted safety for use under medical supervision. Psychedelics that are currently Schedule I include LSD, psilocybin and psilocin, mescaline (from peyote cacti), DMT, MDA, MDMA, and DOM. Of all of these, MDMA, MDA, and DOM are the only ones with overdose potential, typically hyperthermia, serotonin syndrome, and high toxicity when taken to excess. Most often, these overdoses are in combination with pre-existing cardiac or other health conditions and polydrug use. Out of all of these, in total, ZERO have significant potential for addiction. The primary addictive aspect to them resides in the euphoric feelings often experienced, leading more to behavioral addiction to stimulating and empathogenic properties. In fact, most of these psychedelics have overwhelming evidence backing their efficacy in treating addiction, as well as PTSD, major depression, and other behavioral health conditions. A stark lack of evidence regarding potential addiction and harm remains part of the imbalanced approach to scheduling such substances.

We'll get more into this in our final chapter, but I want you to recall the three defining characteristics of Schedule I controlled substances. They are supposed to be highly addictive, have no medical use in treatment, and lack accepted safety for use under medical supervision. None of these meet those criteria--none. Yet consider OxyContin, the harbinger of the opioid epidemic--it's a Schedule II. It's horrendously addictive, constantly abused, and the mismanagement of its use and over-prescription has led to overdose, heroin addiction, fentanyl contamination and death, and lawsuits in the billions. It is, however, less dangerous in the eyes of the US Government in 2024 than psilocybin mushrooms. Benzodiazepines like Xanax, lorazepam, and diazepam are also highly addictive. They pose risk not only for dependence, but the associated withdrawal symptoms can be fatal if stopped suddenly. While working in a detox center, we were able to provide subacute services for withdrawal management without hospital oversight--unless it was alcohol above .25 BAC or benzodiazepines--they had to be under medical supervision due to the risk of seizures, psychosis, and cardiac considerations. They are Schedule IV, a long way from the apparently very harmful mushrooms and cacti hallucinogens.

Historically, psychedelics have been used in ceremonial and cultural traditions as a means of opening the mind, seeing problems from a new perspective, resolving conflicts, and accessing higher forms of governance for the self and community groups. Individually, psychedelics can be used recreationally, though typically the most effective utilization has everything to do with set (what your intention is as you experience the effects) and setting (the context and environment in which you are engaging in the experience). In research on perceptions and intentions of use, individuals engaging in psychedelic experiences report a difference in dose as a primary marker for distinguishing between therapeutic, ongoing support, and recreational use. For some, micro dosing is more effective at managing anxiety, depression, poor self-image, and typically common feelings of being human. Psilocybin, one of the most common and effective plant medicines, increases serotonin activity in several brain regions, particularly those associated with mood regulation (e.g., the prefrontal cortex), emotional processing (e.g., the amygdala), and perception and cognition (e.g., the visual cortex). This boost in serotonin activity contributes to feelings of euphoria, well-being, and altered perception in higher doses, and in micro doses, improves mood and outlook while you remain functional in day-to-day activities. You can imagine the variety of human contexts in which this would be beneficial--whether in small doses to support healthy activity, or at a much higher dose to provide the opportunity to step outside your human condition for a journey outside your daily life.

Clinically, psychedelics were initially being used (LSD in particular) as an approach to cognitive augmentation. Co-opted into research to harm or control others, the fear over psychosis and jumping off the roof thinking one could fly combined into a nasty perception that "tripping acid" was synonymous with schizophrenic madness. This is unfortunate, to say the least, as LSD, ketamine, psilocybin, DMT, and MDMA have incredible secrets to reveal--and we're only now beginning to see the light in attempts to legalize clinical use for these substances state by state. The issue with improperly classified substances like psychedelics extends in two directions, both of which cause undue harm to American citizens.

First, Schedule I controlled substances are defined as having no medical use or sufficient clinical safety--making research and development underfunded, obscenely wrought with barriers to studying effects, and in many cases, illegal to research at all with human beings. Even with clinical research approval, the lack of understanding regarding the use of psychedelics creates issues in quality research. Double-blind experiments, often heralded as the gold-standard for research, are absolutely inappropriate for measuring the efficacy of a treatment that is centered on the participant having full understanding of the substance, their own intentions and internal state, and the idiosyncratic processing during and after facilitated treatment. If anything, a double-blind experiment with psychedelics presents undue distress and harm to participants without contributing to the field of research as intended.

The second issue with inaccurately classified substances is the criminalization of individuals for which these substances are healing and helpful. For example, one patient in my research described the hardship of relocating from a psilocybin-friendly state to one that was not--and knowing his mushroom capsules would result in citation and prison should he be searched. Psilocybin in a guided, facilitated setting had allowed him to crawl out of alcoholism, repair his marriage, and address the issues of abuse he'd experienced as a young man at the hands of his stepfather. He took small micro doses periodically throughout the year, effectively eliminating his dependence on benzodiazepines (he'd been overprescribed anti-anxiety medications for nearly ten years). A state line meant prison or not prison, criminal charge or no criminal charge. The lifelong consequences of Schedule I possession cannot be underscored enough: and yet psilocybin has no real risk of harm and abuse.

Clinical use is growing, slowly, in states recognizing the travesty of how misunderstood psychedelics are and the potential they reveal in treating chronic, treatment-resistant disease and disorders. Psychedelic-assisted therapy, or PAT, is becoming a rallying cry for treating people that have utterly lost hope they can ever find healing or recovery.

Modern use of psychedelics is largely predicated on the perceptions surrounding them. For researchers and clinicians, psychedelic experimentation to address treatment resistant compulsive disorders and PTSD is limited to legal and clinical-use-only substances. In the United States, policies regarding psychedelics are still largely informed by anti-drug campaigns lingering from the massive "war on drugs" waged in the 1960's, with little distinguishing language in classification and scheduling between crack cocaine, heroin, and lysergic acid diethylamide (LSD). Currently, limited laboratory use of methylenedioxymethamphetamine (MDMA) is permitted on human subjects, yet again, the clinical studies are rife with issues due to the nature of the setting being unconducive to the process being observed. Psilocybin has additional limited and ostensibly experimental use in select clinics in very few states. This complicated relationship between science and policy remains a point of contention for researchers and clinicians viewing psychedelics as neurotechnologies and agents of change. Ketamine, an

NMDA receptor antagonist, has been used historically for decades as an anesthetic in higher dosages, which contributes to its legal status. In lower doses, however, it alters mood, consciousness, and cognition, providing an opportunity to complement traditional therapeutic sessions. Ketamine is legal in the United States for use in psychotherapy through licensed providers. Taken in lozenge form or intravenously, ketamine-based psychedelic-assisted psychotherapy is increasingly more available and utilized for individuals with diagnoses of PTSD, MDD, OCD, addiction, and other treatment-resistant diagnoses. One issue with ketamine in PAT is that it is dissociative--for those wanting to engage in catharsis, like that in ayahuasca traditions, it may be emotionally limiting. While the USA still struggles with adequate research into psychedelics, other countries are less limited. Brazil, for example, remains a liberal location for experimental, therapeutic, and shamanistic application of psychedelics, both synthetic and naturally occurring.

Use of PAT to address known and challenging symptoms of developmental or acute trauma-related symptomatology is experiencing increasingly wider use as trauma literature gains popular acceptance and interest, but it remains woefully limited. Trauma-related sequelae have pervasive and cumulative effects on individual health and well-being, ranging from seemingly innocuous startle responses to severe and chronic comorbidities unresponsive to treatment and deleterious to daily functioning. With great significance, interpersonal violence augments internal schemas and memory processes into maladaptive response patterns to the world and those within it. This limits the efficacy of traditional psychotherapy to the diminished capability of the individual's internal mechanisms for managing memories and triggers. Persistent painful memory recall, intrusive thoughts, somatic triggers, sleep disruption, affective dysregulation, and self-medicating behaviors are common in individuals seeking different approaches to treatment, particularly PAT. Ketamine use in PAT has revealed intriguing outcomes that often go unnoticed, unreported, and uncelebrated. Some ketamine-based PAT studies highlight dramatic and immediate shifts in

PTSD symptomology, at times with 80% of patients no longer meeting criteria for the diagnosis one year post-therapy. Evidence that these results are attributable to neural plasticity initiated through ketamine treatment makes the sustainability of such remarkable recovery a point of interest for researchers, clinicians, and patients.

The mechanisms for ketamine's generative influence on brain function align with the degenerative impact of trauma-related sequelae. Traumatic events diminish amygdala and hippocampal volume, are known to produce dendritic clipping or neuronal atrophy, and increased cortisol production in the prenatal environment influences telomere length in offspring. These neurobiological impacts present behaviorally in affective regulation, cognitive flexibility, working memory, and compulsive behaviors. Managing these internalized and externalized behaviors is often the motivation behind treatment seeking later in life. Ketamine very clearly provides rapid and sustained relief of PTSD and depression-related symptoms. Whether applied orally or intravenously, ketamine promotes synaptogenesis in the medial frontal cortex (MFC) and hippocampus. Within a day, just a single dose shows increased dendritic spine density within the MFC, and in some studies, begins to reverse stress-associated dendritic spine loss. This generative neural plasticity, or dendritic growth at synaptic terminals, creates new connections that influence behavior. Recent research in animal experiments show ketamine's glutamate-evoked dendritic spinogenesis at 50% within two hours of administration; the mice within the study also displayed elimination of hyper locomotion (anxiety movements) and decreased depressive symptom behaviors.

Ketamine, along with other psychedelics, is increasingly believed to act with neuroprotective effects via the modulation of microglial cytokine release. From a neurochemical perspective, microglia are noteworthy players in addressing the inflammation associated with chronic and unpredictable stress. Acting as neuroinflammatory neuronal modulators, these agents in the central nervous system (CNS) are, among other roles, the dendritic and neuronal landscapers snipping, tending, and generating synaptic growth. Microglia are increasingly viewed as an effective pharmacological target for mood disorders, traumatic brain injury (TBI), schizophrenia, and other neurodegenerative diseases such as dementia. Additional mechanisms of action for ketamine include its ability to incite perineuronal net disassembly. This promotes the development of neurons, synaptic plasticity, and through modulation of microglial cell activity, promotes homeostasis, or the balance between excitation and inhibition.

While the idea of using psychoactive substances for the treatment of addiction may seem counterproductive, it is critical to understand the difference between illicit drug misuse, the nature of addiction, and the related compulsion behaviors often connected with stress and trauma. Addiction is as complex as PAT efficacy and use, making the partnership of the two a point of great promise. For example, addiction is generally defined and understood to be a result of vulnerable brains with repeated exposure to reinforcing substances. This creates epigenetic changes and neural pathway alterations that lead to interruptions in motivation, executive function and control, and reward pathways. Typical treatments for substance use disorders (SUDs) include three pharmacological approaches: agonist replacement, amelioration or masking of withdrawal symptoms, and limiting the positive or rewarding aspects of addictive substances. These all target receptors and/or enzymes of the drugs, but not the many shared pathways beneath the surface of drug use that create such stable and chronic pathway modifications.

These pathway modifications for individuals with SUD are not isolated to simple drug abuse and eventual addiction. Substance use disorder is consistently included in trauma-related sequelae impacting adolescent and young adult health and wellbeing. Individuals with histories of acute and sustained, chronic and severe, intermittent and even unpredictable traumatic stress are significantly more likely to develop SUDs, for reasons ranging from altered reward pathways to self-medicating avoidance. Often included are risk-taking behaviors and issues of inhibitory control.

Addiction and compulsion share goal-directed behavioral circuitry, partially controlled by the release of dopamine (DA) from the ventral tegmental area (VTA) into the prefrontal cortex (PFC) and the hotbed of reward activity in the nucleus accumbens (NAc). This mesocorticolimbic system is responsible for the pursuit and reinforcement of reward behaviors, which, if repeatedly met with unnaturally increased levels of DA release (such as through drug abuse), can create powerful and overwhelming drives disrupting homeostasis. In short, repeated disruption of reward, fear, and motivational pathways fundamentally alters neurological responses. This diminishes the PFC's authority over behavior and affective regulation, replacing top-down thought with the bottom-up pursuit of immediate drug reward. Much like trauma reduces gray matter in the PFC, reward-system pathway dysfunction from addiction shows similar structural correlates, with atrophy in dendritic spine density and cortical volume. Compulsion is addiction's neighbor. The nucleus accumbens is increasingly the purported seat of obsessive compulsive disorder (OCD), anorexia nervosa, and bulimia, all diseases rooted in a drive to repeat behaviors no matter how harmful or debilitating they might be. Despite its extreme mortality rate, anorexia nervosa is rarely understood as a mesocorticolimbic dysfunction, but this compulsively-natured diagnosis is increasingly connected to reward system dysfunction. Deep brain stimulation (DBS) targeting areas around the NAc has shown some efficacy in treating anorexia, with decreased symptoms of anxiety and depression, and increased BMI.

Noteworthy are the mesocorticolimbic features of anorexia nervosa: disturbance in connecting stimuli and reward appropriately, diminished awareness of homeostatic triggers, and other disruptions in reward and habit circuitry. In addition to the NAc, the subcallosal cingulate cortex (SCC) is also implicated in DBS to treat OCD, depression, and anorexia nervosa. Increasingly, the complexities of compulsion behaviors, trauma, and the psychopathology underpinning SUD have shared neurodevelopmental reward and regulatory system dysfunction. Disrupted development and altered neurological pathways are complex challenges that are well-met with the equally complex interaction of psychedelics. Psychedelics work in dynamic interaction with human neurobiology, with efficacy in treating TR disorders of compulsion and addiction attributed to everything from psychosocial models to neuroinflammatory mechanisms, molecular action, and massive effects on neural connectivity and activity.

Studies abound on the efficacy of single-dose or single-session psychedelic-assisted therapy for such diagnoses. Overwhelmingly, research reveals that psychedelic-assisted therapy for SUD produces long-lasting abstinence, diminished guilt and shame associated with addiction, improved health and wellbeing both immediately and over time, and a stronger sense of agency and self-efficacy. Experiences also improved connection to loved ones previously estranged through addiction patterns and behaviors. Similar to trauma-informed treatment of behaviors and diagnoses, modern research supports PAT as an effective approach to addressing the underlying structures of addiction and compulsion. An additional category outside the developmental aspects of trauma-related sequelae is the overwhelming impact of a chronic or terminal diagnosis and/or acknowledgement of rapidly approaching death. For individuals with terminal diagnoses, such as late-stage cancer, advanced HIV/AIDS in the 1980's and 1990's, or other fatal diseases in advanced states, the approach of death and nothingness presents a terrifying and complex experience. Death that is unavoidable significantly alters and disrupts even the most normatively developed individuals with adequate coping skills. Issues of abject loss of life and ability to be with loved ones, feelings of fear over what happens next, religious and anti-religious beliefs about death and dying, and deep-seated worry that bodily and cognitive functioning will deteriorate sharply are all attributes of individuals seeking support following such diagnoses. In studies assessing PAT outcomes for individuals with advanced terminal illness, pre- and post-treatment assessments were conducted to assess depression, anxiety, acceptance of death, and overall wellbeing. Individuals reported heightened symptoms in the pre-treatment assessments of diminished self-worth, disconnection from the healthy, active world, deep resentment at their diagnosis, depression, anxiety, fear over death, and suicidality. Following treatment weeks, months, and years after diagnosis right up to recovery or death, individuals reported significant reduction in negative symptoms and increased positive feelings. These included acceptance and love for the world, a strong sense of connection and purpose related to God or a creative force, hope for what lies after death and dying, and peace regarding their diagnosis.

For individuals in the HIV/AIDS epidemic of the late 20th Century, an added trauma to a then-fatal diagnosis was the social isolation, judgment, and utter lack of adequate and equitable medical care that compounded feelings of anger, dissociation, major depression, and intense fear. Following PAT, individuals previously experiencing resentment, pain, hostility, fear, and isolation reported overwhelming forgiveness, a desire to make the most of their remaining time, a deep connection to a universal good, and feelings of acceptance and peace regarding their diagnosis and death. This changed perspective sustained over time, with consistent reporting of positive and peaceful affect and feelings related to the diagnosis for years following their PAT sessions.

It is unsurprising to me that our society as a whole has struggled with identity and health. We have been misled by political leaders changing public policy and legislation to suit their own agendas, and been on the receiving end of abusive and profoundly greedy pharmaceutical and insurance industries masquerading as providers of care. When you struggle with knowing who to trust and how to recover, you look for answers wherever they seem to be--and industries profiting on consumption and illness will always have answers. For an enormous number of people, traditional short-term therapy, exercise, proper nutrition, and overall social and metabolic health are ingredients that support self-regulation, healthy habits, and overall wellbeing. For those with chronic issues related to traumatic events, options are limited to masking, managing, and avoiding. Often pharmaceuticals mask and manage symptoms, while long-term therapy and treatment provides *support* for symptoms; so ultimately, many people opt to avoid the deep-seated origins of dysregulation and poor health with distraction and avoidance in the form of "treatment".

Distraction is a very, very easy pit to fall into. When managing symptoms rather than causes, distraction is the name of the game. Trauma-induced chronic disorders are often lifelong, cumulative, complex, and provide a host of opportunities for diagnoses that could very well be contributing aspects to multifaceted treatment resistance. The desire of symptom masking and avoidance can readily be pushed to the side when compared to the neurogenerative, sustainable, internally navigated shift in perception offered by PAT.

One helpful example is laser eye surgery. Laser eye surgery, commonly known as LASIK (Laser-Assisted In Situ Keratomileusis), is a procedure designed to correct vision problems such as nearsightedness, farsightedness, and astigmatism. During the surgery, a specialized laser reshapes the cornea--the clear, dome-shaped surface at the front of the eye--to allow light to focus more accurately onto the retina, resulting in clearer vision. The process begins with the creation of a thin flap in the cornea, which is gently lifted to expose the underlying tissue. The laser then precisely removes microscopic amounts of tissue based on the patient's prescription before the flap is repositioned. Without LASIK, people with vision issues can use glasses and contacts. The visual apparatus does not change, but its deficiencies are managed and mitigated with tools you can use to shore up the dysfunction. LASIK goes into the mechanism itself and augments it, resulting in the visual apparatus being reset to a standard, healthy operation, capable of receiving information and processing it appropriately.

Psychedelic-assisted therapy is the LASIK of trauma treatment. Psychedelics enter the mechanism of the brain, right down to the dendritic spikes, neural connections, and serotonergic and dopaminergic systems, opening up the individual in two ways. First, measurable improvement in neuronal health and function; second, a perceptual opening in which memory and identity are unseated from reality.

Neuronal health begins, as we learned already, in utero. Telomere shortening in stressed babies contributes to earlier cellular degeneration, and neurons react and adapt to toxic stress as much as possible before their demise. For those with trauma-related dysfunction, dendritic spine sparsity and synaptic fragility are common. It means, generally, that your information connectors and processors in your brain are poorly equipped to deal with oxidative stress, aging, and inflammation. Studies increasingly reveal that psychedelics have promising effects on neurogenesis--the process of forming new neurons--particularly in the context of mental health treatments. Research into psychedelic effects on the brain has shown that it can promote neuroplasticity and enhance neuronal growth in regions such as the hippocampus, a brain area crucial for learning, memory, and emotion regulation. Psychedelics like ketamine and psilocybin are also known for their rapid-acting antidepressant properties, especially in cases of treatment-resistant depression. Studies indicate that psychedelic substances' ability to stimulate neurogenesis and synaptogenesis (the formation of new synapses) in the hippocampus and prefrontal cortex contributes to these antidepressant effects. Ketamine also increases levels of Brain-Derived Neurotrophic Factor (BDNF), a key protein involved in the growth, survival, and differentiation of neurons. Higher BDNF levels are associated with improved neurogenesis and synaptic plasticity. It also activates the mTOR (mechanistic target of rapamycin) signaling pathway, which is critical for protein synthesis and the formation of new synaptic connections. This pathway plays a role in ketamine's ability to rapidly reverse synaptic deficits caused by stress or depression. It restores dendritic spine density, and by modulating the glutamatergic system (specifically NMDA receptors), ketamine enhances the release of glutamate, which also promotes synaptic growth and neurogenesis. Ketamine, while synthetic, is not alone in neurogenesis and cellular health in the world of psychedelics.

Psilocybin, our funky fungi, is converted in the body to psilocin, which binds to 5-HT2A receptors--a type of serotonin receptor found throughout the brain, especially in the prefrontal cortex and hippocampus. This receptor activation promotes intracellular signaling pathways that enhance neuronal growth and plasticity. It also does what ketamine does--psilocybin boosts the levels of BDNF, that protein crucial for the survival, growth, and differentiation of new neurons and synapses. Higher BDNF levels, you'll recall, are associated with the formation of new neurons, improved synaptic plasticity, and cognitive improvements in memory, learning, and emotional regulation. It also has been shown to reduce neuroinflammation and oxidative stress, both of which can hinder neurogenesis. In making the brain healthier through reducing inflammation, psilocybin creates a more favorable environment for new neurons to grow and survive. Reduction of inflammation in the brain has enormous implications for a variety of brain diseases, including Alzheimer's/dementia. In fMRIs of humans, psilocybin is shown to improve connections between different brain regions, reflecting underlying neurogenesis and neuroplasticity.

These impacts are absolutely mirrored in the other psychedelics used in therapeutic contexts, so much so, that it becomes redundant to continue listing the impacts. Summarily, psychedelics grow your brain connections, with particular care and support for the areas impacted and damaged by traumatic experiences. Psychedelic-assisted therapy is an epidemiological miracle for resolving treatment-resistant disorders. It's time we start seeing it as such.

8

A STUDY IN HOPE

From 2022-2024, I embarked on a research journey that had been percolating for quite some time within my heart. I have always had a curiosity and desire to understand healing, as I do feel my own life has been a reflection of the powerful hand healing has in guiding an individual into a positive life experience. Healing carries a lot of nuance: it's generative, multifaceted, and often surprising in how and when it happens. For some people, healing comes from real consequences to behaviors that are harmful, and for others, healing comes quietly and subversively through a series of increasingly positive moments acting as a salve to historic injuries internal and externally inflicted. For my doctoral work, I was interested in healing from the perspective of dramatic change--from broken and struggling to healed and motivated. From being mired in destructive patterns and subject to the whims of impulse and trauma-related sequelae to careful, self-initiated progress to health and wellbeing. The portion of individuals that simply cannot access health and wellbeing naturally became the population I wanted to study--if consistency is the hidden structure of dramatic change, what happens if you are unable to be consistent? Can you ever change?

Psychedelic-assisted therapy (PAT) became increasingly salient in my research once I understood, in even a limited fashion, the impact traumatic stress has on the neurobiology of developing human beings. Environmental factors influencing genetic expression (epigenetics), the wealth of information on Adverse Childhood Experiences (ACEs), a personalized understanding of ADD/ADHD as a potential symptom of trauma-influenced development, and of course working with individuals struggling to find any sort of consistent, sustained approach to behavior change...psychedelics wove itself almost invisibly until the last moment. It popped its head up periodically--in talking with individuals that had life-changing experiences, seeing remarkable improvements in historically destroyed human beings, and of course, in reading and listening to both novice and expert perspectives on psychedelics through history to today. The fact that the serotonergic and dopaminergic systems assaulted and stunted by stress events are the very systems in which psychedelics are neurodevelopmental and generative could not be ignored.

During the course of this research, I interviewed patients that had completed psychedelic-assisted therapy at least six months prior to the study. I was often surprised and delighted throughout the process by the incredible honesty from each patient. Their self-awareness, voluntary vulnerability, and truly genuine dialogue was transformative in so many ways. The intention of the study was to identify how our regulatory system changes with PAT, but not strictly from the neurobiological side or the behavioral side. I wanted to know what the patients thought about their own histories, why they engaged in PAT, and how they would describe the change in their life following it. The proof of the pudding is in the eating, so to speak--I already knew PAT produces neurogenesis in the areas impacted by trauma, and that those very areas impacted were related to dysregulation and chronic disease. Now, I wanted to see how patients' lives actually played out--were they better? Did they relapse? Most importantly, *did they believe their lives were significantly improved?* Were they able, at long last, to actually engage in consistent, sustainable change, previously inaccessible with their hijacked, trauma-influenced neural circuitry?

The question hardly seems worthy of mystery, and if you've stayed this long with me in this work, you can likely guess the answer.

Yes. Without a doubt, yes.

The variety of chronic, treatment-resistant disorders was particularly intriguing. For almost all patients, childhood adversity was rooted in parental or familial dysfunction. There were gendered aspects to this. Largely, the women I interviewed that had familial hardship related stories of abuse, neglect, and psychological manipulation by their mothers. Often, the mothers displayed, upon retrospect for the patients, mood disorders or other psychopathy that significantly augmented their child's perception of self-worth. Mothers who would become irate when their daughters enjoyed time not spent with them; mothers that would set a moving target for what was acceptable and deserving of love; mothers that would punish and isolate their daughters for maturing or seeking friendship outside their dyad. Summarily, mothers that saw their daughter as an extension of self, a source of envy, or an object for abuse or consumption. One beautiful, brilliant young woman described growing up in a home with a mother that truly saw her as a possession, almost doll-like. Perfection was expected but never defined; happiness *had* to be derived from mother, and yet her mother was chronically unhappy and unable to manage her own emotions, thoughts, and behaviors. Any happiness from outside of her mother's provision was rapidly hidden, dismissed, avoided, and stuffed down--because it was so deeply upsetting to her mom. Imagine what this does to a little girl, developing an identity in an environment that repeatedly reinforces that she is bad for being happy and making her own choices. She hurts the most important person in her life when she has independent thoughts, ideas, or behaviors. If she makes decisions, they are always wrong. She must wait for her mother's mood in the moment to determine her choices, effectively moving the locus of control to her mother. She learns to avoid telling the truth, especially about her thoughts and feelings, because they will be offensive, hurtful, and selfish. She develops an identity that is rooted in shame, as the lack of consistent expectations means chronic tiptoes through egg shells. She looks for others to inform her choices, as the control over such choices is always in the hands of her mother. Perhaps most critical and most heartbreaking is the identity of the little girl being formed with the understanding, since birth, that her needs and wants are a distant consideration, if at all, and that she causes unhappiness by simply existing. This little girl grows up knowing--not believing, but *knowing*--that she is not worthy of sacrifice by another. Love is transactional: how happy you make a person, at the cost of dissolving yourself, is how you obtain love.

Consider the implications this has for the young adult. Even if she never sees her mother once she steps out the door at 18, the young woman has an identity that is only a shadow of a self. If she wants love, she has to find someone who wants to consume her, because that is the understanding she has of love. If she has to make a decision, she looks to someone outside herself, because she has learned she cannot trust her own choices. When she has a need or a want, she will stuff it away, avoid it, and ignore it--because those are not acceptable things to have. They hurt the people that love you. When she meets a new person, she mirrors their identity, because that is how she's learned to navigate dyads: mirror the other person, because if you show them yourself, they will be angry and offended. This can't last forever, though--eventually, the dissonance between what she must pretend to be and what really is becomes overwhelming, and she begins to transform suppressed, normative desires for autonomy and self-efficacy into other aspects of control that are less visible. She begins to binge on food and alcohol, and then forces herself to vomit. She seeks out relationships that are extractive, in which she feels familiar in patterns of giving over herself, her thoughts, her choices, and her identity to consuming, unhealthy partners that seek to control her. The cry for relief finds a way, though. It leaks out, in the vomiting with bulimia, the uncontrollable sobbing at work, the constant seeking of abusive relationships. These are links in the chain of events from childhood trauma that wire perception and behaviors in a tragic display of symptoms echoing unmet needs for safety and love. The young woman I was talking to seemed so different and distant from the young woman she was describing--an unstable alcoholic with an eating disorder, dysfunctional at work, and with a chronic history of abusive relationships. The woman being described had tried all sorts of therapy--cognitive-behavioral, somatic, traditional talk therapy, even eye-movement desensitization therapy (EMDR). This treatment involves guiding the patient through specific eye movements while they recall distressing events, and has shown effectiveness for many people with trauma. Therapy like EMDR helps the brain reprocess these memories, reducing their emotional impact. Sometimes these things would help for a bit, or even a little longer than a bit, and then *wham.* Symptoms return with aggression. She believed at this point in her adult life that her tight chest, anxious stomach, chronic unhealthiness, and uncontrollable emotions were firmly hers, forever. She tried to kill herself, for what was not the first time in her young life. Yet here she sat, across from me, nowhere *near* who she was describing.

This woman before me was insightful, confident, well-spoken, naturally beautiful, obviously healthy, and completed an advanced degree while working professionally. She was happily and healthily remarried, could have a treat without binging, rarely touched alcohol, had healthy boundaries with her mother, and had navigated, through all of this, significant and potentially fatal health issues. Here's the shocker--she had a lifetime of chronic disorders and disease, failed attempts at all types of therapy and medications...yet her psychedelic-assisted therapy sessions were less than 18 months prior to our discussion.

One of the most remarkable things she shared moved me deeply, and created an accidental lens from which I viewed subsequent patient stories. She described laying, waiting for the effects with her trusted therapist sitting by her side, and remembering the pre-work of opening doors when she saw them in her mind. She saw in this session a little girl, and knew immediately she was meeting herself as a child. Time is suspended with psychedelics, one of the most beautiful aspects to their properties. The necessary dissociation from the current self and state to one where you are untethered to your memories allows you to examine and walk within them, pain-free. You walk around your wagon, your bear, your trauma.

I asked what she said to herself. Her response was, "I held her hand, and I told her she was enough. That she was worthy of love." She then added that after that session, she was able to hold on to that experience almost as a memory from childhood. She met her childhood need for love and acceptance by moving back through time to provide it *herself.* This is beyond remarkable when considering the implications of navigating historic injuries outside of time—you take, suddenly, the position of observer, supporter, listener, and friend. Even, perhaps, the parent the child within never had--and needed. She had the ability to explore, dissect, observe, and navigate her child-self and her needs outside of time and painful memory. Can you imagine the impact of such a concept? Can you imagine being on a balcony, a la Juliet Capulet, observing the machinations of your life from above, viewing what has been, is, and will be? Furthermore, can you envisage being able to *remember* such events, with a keen eye and level mind, in order to incorporate your understanding into life from that moment on? The ability to return to a point in time where a pivotal moment occurred, and be able to guide the trajectory of your destiny in an intentional way...it's almost jumping timelines at the junction point of the most impactful hurt. This is inconceivable, typically, and yet it's not science fiction. It's real, and accessible, and worth every effort to explore.

This remarkable opening of the mind and walkabout through its halls is a common feature in many PAT sessions--doors, hallways, rooms, and staircases are described by patients in many studies. The importance of trust in your clinician or facilitator cannot be underscored enough, and all patients in my study emphasized the importance of that relationship in addition to safety in the setting itself. You are deeply vulnerable when on a journey of the mind, particularly when the mind you are navigating is holding secrets long hidden and festering in closed rooms from long ago.

The surprising nature of walking through one's mind was highlighted in one patient's shock at uncovering a trauma she'd previously dismissed as inconsequential. Fully anticipating a journey in which she speaks at length to her mother, a source of profound abuse during her childhood, she did her pre-work and knew to open any doors that may come up. During her session, she opened the door, and to her surprise, it was a boyfriend from when she was a young woman. She shouted his name in surprise, and asked why he was there. His response was, "It's okay, you can let me go. It's okay, it's not your fault, and you can let me go." This boyfriend had been older than her, when she was wild and free. During their relationship, he got an terminal diagnosis, and this terrified her. She'd been living life without tethers, never staying anywhere long enough to love or be loved. She was running from commitment because her entire childhood was shaped by the idea that if you stay somewhere long enough, everything will be destroyed. She came out of her session sobbing. When her clinician asked her, "Who's name was that you shouted?" she came undone. She'd left him, in her twenties, because she had no idea how to care about someone. She was terrified to lose him, so she cut him out of her life before he could leave her. He died shortly after she left. She hadn't realized until that moment that it was precisely then that her drinking worsened, her work got disjointed and sporadic, and her sleeping and eating became dysregulated. When he freed her, from beyond death's wall, she freed herself. The woman I was talking to had undergone PAT after nothing else worked, and was now in her mid-fifties. She works in the medical field, is carefully growing her circle of trusted friends, and is even able to navigate relationships with family members that were historically fractured. I asked her if she had any advice or insights for people regarding PAT. She described it as the hardest work she'd ever done. She cautioned against seeing it as a magic pill or instant fix, but confirmed the change in her psyche and self-perception was immediate. The lasting changes, she added, required constant reflection and intentional self-love, which she could do for the first time in her life. She had originally engaged in PAT to help with her alcoholism--and came out with a completely changed perspective on her life. This was in addition to no longer experiencing dependence physically or behaviorally on alcohol.

It was fascinating to learn about the experiences *during* PAT sessions. Patients reported both similar and distinct experiences, with sensory and emotional experiences often coalescing in synesthesia-like sensations. Interestingly, men and women reported many similar experiences with the exception of visual phenomena. Many women described more concrete visuals, such as walking through hallways and encountering people, while many men described geometric shapes, often carrying with them impressions of emotions and feelings that took on colored hues. Women reported a sense of compassion, a generative creative force, and a sense that the universe and all within it are bound by a tremendous love. Men reported similarly, though often made comments like, "I could see the structure behind things, the patterns in life that repeated outside of time". Not all, but most of the men shared a sense of deific encounter, moving the needle of faith and spirituality further toward belief and comfort that their existence was part of something greater. All patients, regardless of gender, age, or circumstance, genuinely and emphatically reported that the dissolution of their ego during their treatment led to an increased ability to access critical epiphanies, in which they were able to see historic traumas objectively and compassionately.

In each patient's story, both unique and shared themes emerged. The motivation for pursuing PAT was rooted in past trauma and related sequelae that followed, largely from childhood and familial dysfunction. Traumatic caregiver experiences and marriages, chronic pain and depression, sudden and toxically stressful physical diagnosis, exhaustion over failed treatments and loss of identity through disability featured prominently in many cases. Long-term bulimia and disordered eating, deep frustration and shame about failed treatment for chronic conditions, severe PTSD, major depression, suicidal ideation and attempts, unresolved emotional pain, and substance abuse were all drivers for patients seeking PAT. For almost half the patients, chronic *physical* issues exacerbated and compounded chronic behavioral health issues, creating a vicious cycle of failure in recovering from any condition. The uncontrollable nature of physical decline, injury, or disability was a strong component of hopelessness felt as treatment after treatment fell short.

Completing between three and twelve PAT sessions, patients described differences between their own sessions based upon previous journey work and reflection. For one young man struggling with early-onset musculoskeletal degeneration, the loss of identity as an athlete separated him significantly from social groups he'd been a part of with that persona. When he had extensive surgery and felt "broken", he lost relationships he'd believed were based on more than shared activities. The spiraling of chronic depression since childhood, lost social connections, decimated identity, and the finality of physical disability left him in a world of monochromatic listlessness and hardship. He reported that, despite being very young and creative, he was unable to find joy in anything. The world, as he put it, "dragged me down". He was mentally exhausted from previous treatments, including traditional counseling, meditation, physical therapy, prescriptions for fluoxetine and duloxetine, and the terror of being utterly unable to control his physical decline. His three PAT sessions began with a sense of more control over the experience, and a sense that the sessions were manageable. His third session, however, he described as chaotic and intense, including flinging between structured and nonsensical in his sensory experience. He began to process life from outside his body and mind, moving through textural and visual imagery that symbolized personal realizations. "I had a spiritual connection," he reported. His connection to his faith was strengthened, and he felt he had experienced contact with a larger design and purpose than his physically limited state had provided previously. After lifelong depression, poor boundaries and unhealthy relationships, and still with chronic pain and recovery needs, this young man reported that experiences now had "a new shine". He reported improved management of emotions, an ability to manage historically destructive stressors with grace and compassion, refined boundary setting improving his personal relationships, and dramatically enhanced discernment over relationships, meditative practice, and intentional joy.

When it comes to healing, the road often winds through the most unexpected places. It is a journey that is both personal and universal, and the words we use to describe it are charged with significance: trauma, self-worth, release, boundaries. These words, so simple on the surface, carry the weight of years, sometimes even generations, of unprocessed pain and undigested grief. They are the markers of our struggles, the signposts pointing toward the places where we've been broken, where we've been taught that we are less than enough.

For many, trauma is the beginning of the story. Whether it's the wounds inflicted by an abusive parent, the neglect of a caregiver, or the loss of safety at the hands of someone who should have protected us, trauma becomes the lens through which we see ourselves and the world. "I am not good enough." "I am not worthy of love." These beliefs take root deep within us, shaping how we show up in relationships, in our work, in our very sense of identity.

Yet, what if the stories we tell ourselves about who we are not the whole truth? What if the trauma that once seemed like an unmovable wall could be softened, reframed, and even healed? That's where healing begins to take on a different meaning. The first step is understanding that trauma, though it shapes us, does not necessarily define us. It's through moments of catharsis that we begin to free ourselves from the grip of these old beliefs. The words we use matter here: release, transformation, surrender.

In psychedelic-assisted therapy (PAT), those moments of release become tangible, and are often unexpected. I've listened to countless stories of patients confronting their pasts, only to be taken aback by what emerges from within. Consider the woman who came to therapy hoping to address her strained relationship with her mother, only to have her first session bring up unresolved guilt about a past relationship. Or the young man, dealing with chronic pain and depression, encountering a sense of divine connection during his session. Or the young woman, giving her child-self the love she needed as a child. These were not just memories or experiences—they were insights, deeply personal revelations that changed how they saw themselves and their lives. They were revelations inaccessible through the fog of pain.

The beauty of PAT lies not just in the moments of release but in the integration of those insights into daily life. It's here that the real work begins: setting boundaries, redefining relationships, and shifting self-perception. In the aftermath of these sessions, patients find themselves more able to establish healthy boundaries, to say no to what no longer serves them, to embrace their own self-worth without the need for external validation. Words like "equanimity," "self-compassion," and "mindfulness" start to make sense in a way they never did before. They become practices, not just concepts.

We must not forget the role of self-compassion in this process. For those who have spent years struggling with shame, guilt, and self-hate, learning to show kindness to oneself is nothing short of revolutionary. I've spoken to women who had spent their lives trapped in toxic relationships, constantly giving of themselves until there was nothing left. The first time they learned to hold space for their own needs, to acknowledge their own pain without judgment, was like a light being turned on in a room that had been dark for years.

Of course, change doesn't happen overnight. Healing is not a magic fix, it's not a one-time event. It is an ongoing process of growth, of peeling back layers, and finding new ways to show up in the world. That's where the words "transformation" and "curiosity" come into play. Transformation happens when we begin to see ourselves with new eyes, when we let go of old beliefs and make room for something greater. It's about understanding that we are not defined by our past mistakes or traumas, but by our capacity for growth, love, and change.

This is why psychedelic therapy holds such promise—not because it offers a quick fix, but because it invites us into a deeper understanding of who we are. It gives us the space to reframe our stories, to release the weight of old wounds, and to rebuild our sense of self from a place of self-compassion and courage. The journey of healing is long, but it's worth it. Each step forward is a testament to the resilience of the human spirit, and each moment of release is a reminder that we are always capable of more—more love, more peace, and more wholeness. It's in the moments of deep introspection, the release of old emotional patterns, and the integration of new insights that we reclaim our true selves. Healing is not just about recovery; it's about becoming whole again.

It's fitting to close this chapter with a number of the responses to my question during my study. This question was, "What is something about psychedelic-assisted therapy that you think is important for people to know?" The answers were thoughtful, and consistent in their messaging.

"This is not easy. But it is worth it in every way."

"I'd tell people who think psychedelics are only drugs, or are dangerous and should still be outlawed...to take a look at alcohol. Take a look at how destructive that is. I want people to know that alcohol destroyed my childhood, destroyed me as a young adult. I want everyone to know that I am healthy and happy because psychedelics helped me heal from the pain alcohol caused my entire life."

"I'd say to go in as a beginner. Like you want to set yourself toward being open, and being humble. Then it helps you make the choices you weren't even able to see, let alone choose."

"I would tell people that psychedelic therapy was the only thing that worked. That I'd be dead without it. That's it, that's all--I would be dead if it wasn't for this therapy."

"This was the hardest thing I ever did, and it's also the best thing I've ever done."

Perhaps my favorite response is one that came from the young woman in one of my first interviews. She had never had a moment, since her earliest memories, where she felt she had any say in her own life. She had lived with abuse, illness, eating disorders, major depression, PTSD, severe anxiety, panic attacks, alcoholism, suicide attempts, and exploitative relationships for her *entire life*. She was never permitted joy, or safety, or autonomy. When I asked her what psychedelic-assisted therapy meant to her, and what she'd like people to know about her experience, her reply was simple.

"Now I have hope."

9

NEUROGENESIS AND BEHAVIORAL CHANGE

So what really happens to the brains of the people who have undergone PAT for trauma-related chronic disorders? Why is it that a session, or six sessions, all less than five hours in length, can produce such profound, life-changing experiences where traditional therapies have been ineffective?

Neurogenesis, or the development of increasingly robust neural networks and healthy cellular function, is typically a feature of early and adolescent development periods. This is why, as I'm certain you've understood now, childhood traumas have such powerful influence over adult behaviors. Compelling research into early childhood trauma and adolescent critical period plasticity reveals the ability of the human brain to adapt and adjust to new phases of life and development. Adolescent critical period plasticity refers to a phase during adolescence when the brain is particularly sensitive to experiences and environmental influences, making it a period of heightened neuroplasticity, somewhat similar to early developmental phases. Neuroplasticity is the brain's ability to reorganize itself by forming new neural connections. While the brain is always capable of change, certain periods, such as adolescence, are more "plastic" than others, meaning the brain is more capable of adapting in response to experiences.

During adolescence, the brain undergoes significant structural and functional changes, especially in areas associated with higher-order cognitive functions, emotional regulation, and decision-making. The prefrontal cortex, responsible for executive functions like planning and impulse control, matures during this period, and this can have a lasting impact on an individual's cognitive and emotional development. Adolescence is marked by synaptic pruning (the elimination of unused or unnecessary neural connections) and myelination (the process of coating nerve fibers with a fatty substance that enhances signal transmission), both of which contribute to more efficient brain functioning. During this period, the brain is especially responsive to environmental stimuli, such as social interactions, learning experiences, and even stress. Positive influences can enhance cognitive and emotional development, while negative experiences (like trauma or neglect) can have long-lasting effects on mental health and brain function. It's also a critical time for acquiring new skills, particularly complex ones like language, abstract thinking, and emotional regulation. The plasticity of the brain during this time allows adolescents to adapt and learn more easily in response to their environments. Adolescents are also more sensitive to rewards and risks, partly due to changes in the limbic system (the brain region involved in emotions and motivation) and the ongoing maturation of the prefrontal cortex. This can influence behavior, leading to risk-taking tendencies, but also the potential for reward-driven learning. Research has shown with increasing volume that traumatized children respond powerfully to supportive, consistent, loving caregiving during adolescence, with many children "beating the odds" so to speak when it comes to chronic disease later in life. Much of this recovery and neurogenesis is a result of capitalizing on the plasticity of the adolescent period, with co-regulation of emotions and stress response from healthy caregivers having a greater impact than in other developmental periods. To use the operating system analogy once more, intentional and trauma-informed caregiving for children with histories of trauma is like a software patch, fixing existing flaws in the circuitry while the software code is more easily augmented. It requires time and effort to "stick".

Perhaps one of the most challenging aspects of providing such meaningful care is that many children with histories of trauma are subject to further traumas, as their environments stand to remain the same, or worsen. The removal of a starved and neglected child to a foster home where they are beaten does nothing but add complexity and dysfunction to regulatory systems. The removal of a starved and neglected child to a foster home filled with routine and love will do such tremendous good that by adolescence, the opportunity to nearly fully erase the impacts of trauma is not only feasible, but *likely*. These windows of opportunity are not always open, however, and the dual nature of plasticity creates significant vulnerabilities in developing humans, too. This is why so many childhood traumas move into adulthood. Rather than leveraging adolescent critical period plasticity to boost regulatory function and carve out positive identity, many adolescent caregivers are ill-equipped to provide thoughtful, consistent, loving parenting. They may be the very individuals responsible for the traumas, or they may simply be unable to navigate *their own* unresolved hurts, and the cycle of dysfunction perpetuates generationally. Environmental factors like poverty, family illness, and parental separation provide even further risk factors for children aching for structure and safety. Brokenness at home equates to stress overload at school, impacting academic performance, attention, memory, behavior, and peer relationships. The spiral is so easily observed forensically, and yet for many teens and young adults, it's unobservable until it's pulled them to the rockiest bottom possible. And there are so few tools at their disposal to claw their way out.

Executive functioning is one of the most obvious interruptions to self-regulation in adolescents. Manifesting as hyperactivity, fidgeting, daydreaming or spacing out, inability to focus on most tasks, hyper focus on certain tasks, obsessiveness, impulsivity, interrupting others without meaning to, disorganization, loss of time, and emotional lability are all part-and-parcel of being unable to execute tasks. Executive dysfunction looks like ADHD/ADD, OCD, anxiety, and often depression follows suit as serotonergic and dopaminergic systems struggle to maintain equilibrium, creating highs and lows that can be misdiagnosed as bipolar or other mood disorders.

I posit here that psychedelics are capable of creating an alternative, intensive, instant, and short-term period of plasticity mimicking that of adolescence. This is the neurobiological "opening" or "rewiring" that is observed and related by patients. This can provide new approaches to old problems, resolving deficits in care through loving compassion from the self to the self, identifying new reasons why or how something has happened, and looking at internal shame and hurt from the perspective of someone unsaddled with such feelings. They also serve as expanders of consciousness, revealing the massive interconnectedness of life itself. Some people stare at the stars, or the ocean, and see an abyss. They feel small, insignificant, and vulnerable. Psychedelics help dissolve that egocentric idea that the universe is massive and we are vulnerable interlopers hoping to avoid pain but ultimately the focus of the unknown. They teach us that we *are* small, and we *are* vulnerable, and that we are in the magnificent position to be conscious within a fabric that we are innately a part of, and will return to.

When you look at the stars, you have a choice. Observe as a victim, terrified, or stand in awe that the makeup of stars is manifested in *you*.

You matter. You are important. You deserve love.

If our wiring is damaged by harmful events, removing our sense of belonging and value, then every subsequent interaction will be framed by isolation and worthlessness. If we can insert a patch, or install a new OS with updates that revise the circuitry with some effort and guidance, why wouldn't we?

If you have a sick child, you provide them with treatment epidemiologically aligned with their illness. You do not see a feverish child vomit and expect that simply saying they are sick will make them better. Similarly, you do not see a severely dysregulated person and expect a diagnosis of PTSD will make them better. The identification of the *why* behind the symptoms is step one. Step two is providing reasonable access to treatment that addresses the *why*, rather than the *symptoms*.

Treating symptoms is insufficient. Chronic disorders are not a cold. Yet this seems to be our cultural approach in the United States. If the symptoms are bothering you, take something to mask them. That's well and good for temporary hardships and challenges, but hardly scratches the surface of the complexity behind severe dysregulation and the related neurobiological processes.

One population is often overlooked when we consider the vulnerability that accompanies traumatic experiences, and despite our proclamations of honor and support, we continue to facilitate their trauma while refusing to provide them with the most effective intervention we have. We have briefly touched on the trauma associated with military service, and now we will expand that. Our next chapter is all about combat Veterans, and exactly why blocking their access to psychedelic-assisted therapy is the blackest of marks and cruelest dishonor we can bestow upon them.

10

VETERANS

"Thank you for your service." This statement has been the topic of responses ranging from humorous to dark, blithe to polite. My husband's periodic response has been, "just another cog in the war machine", to which I try to remind him that people say it for *themselves*, not him. They aren't sure how to say thank you, or how to recognize and verbalize an abstract concept like volunteering to enter combat, or some visage of it. People say *thank you for your service* like they say *I'm so sorry for your loss.* They are words of social acknowledgement that you have lived something individually experienced yet shared with many, and there is little else to say in that moment. Perhaps it is an honorific, though I'm inclined to believe it's moved to a platitude, through no fault of the utterer, but rather more the continuous insertion of human beings into an industry of war.

Throughout this chapter, I will refer to Veterans with the capital "V" out of respect, and with the note that when discussing complex PTSD, wartime violence, and brotherhood, I am speaking primarily of the men who are a part of significant combat events and special operations units. Bust-the-wire Veterans. This is not to discount or brush aside other men and women who have experienced combat--this is specifically to address the investment and cost required of these men, and the uniquely terrible inability many have to reacclimatize as a result of their service. The selection process for elite soldiers is generally understood to be grueling, as the intention is to skim the cream from the top for use in specialized, hyper-performative contexts. This is performed with a psychological approach that dominates even the physical, in my view from the outside. The intention of basic training and typical service sectors is to dismantle the individual and build up the unit. By the time high performers reach selection courses, many of them are already of the mindset that "regular military" is not their forte--they are aiming for autonomy, intense competition, elite performance, and the ability to have the absolute shit tested out of them to see if they come out on top of the rest of the chaff. This mindset is already one that is deeply embedded with cognitive, emotional, and physical resilience. These men are successful because they are all but *impossible* to break. It works well, because the job of these men is to be unbreakable. The job itself is *designed* to test breaking points. The issue is not this match made in battle, but rather, the utter lack of understanding on how these men return to a world designed for humans that are entirely different creatures. They are returned to social systems with questionable mores and unimpressive authorities that hold structural power over men a thousand times stronger than they are on their best days combined. These authorities can persecute, arrest, prosecute, and demand compliance; the Veterans subject to their whims and edicts have been trained to defy resistance, subvert power structures, and (at times literally) dispatch despots from planet Earth. They come back to power-hungry police, selective enforcement practices, and a world that tells them all they trained for doesn't work here, so either acclimate or go to prison. It's a large part of why many high performing Veterans relocate in rural, remote areas. They don't necessarily want to be lonely--they want to be *left alone*. The difference is important.

Divorce rates for elite forces have been noted as high as 90%, often a result of constant deployment, stress, and mental health issues upon return from combat. It hovers, typically, around 80%. The number of compulsive behaviors and diagnoses for high-performing special operations soldiers is difficult to track. Reporting mental health issues, revealing behavioral health concerns, and seeking support or intervention is far less likely in this population than in civilian contexts. The nature of the job is to move through discomfort and injury--extending to the psychological and emotional. Recall the teeter-totter inside the brain that seeks balancing between inhibition and motivation; consider how dramatic shifts in cortisol could create physiological drives to recalibrate from inordinate stress to feeling good. Factor in psychological isolation when not deployed, loss of purpose, loss of brotherhood, stress in returning to family life, and challenges in sleeping, routine, and re-acclimation to a different way of living all together. This is the perfect storm for dysregulation, and to compound it, the experience of witnessing and participating in violence creates even more intensity in the stress and pleasure imbalance.

While research is limited on elite forces compulsive sexual behavior (CSB), anecdotal and inferential data indicates that CSB is one of the top issues related to marital issues, divorce rates, STDs, and destructive social patterns. This is hardly surprising, when you consider the state of high-performing soldiers psychologically, socially, and physically. The pursuit of sexual activities are typically present in healthy young men, yet the addictive aspects of CSB push that pursuit into the realm of psychological rather than biological. The need for intimacy, love, acceptance, comfort, safety, and pleasure all coalesce in seeking out sexual partnership; the bump to the dopaminergic and serotonergic systems by way of pursuit, conquest, vulnerability, intimacy, and connection are consequently reaffirmed and reinforced with oxytocin dumps during coitus. I connect oxytocin with attachment, based upon the robust body of research on infant-caregiver attachment, emotional co-regulation, and methods of comfort and care in early childhood. In an environment with intense stress and constant danger, a sexual experience can be comforting and offer relief from the constant cortisol onslaught. For a soldier returning home, the absence of danger and imbalance of intensity easily leads to pursuit of comforting, distracting, and thrilling sexual encounters, even if brief and unrepeated. Similarly to compulsive eating, drinking, or thrill seeking, CSB is an attempt to fill an unmet need. It has different consequences, though--even with two consenting adults, issues of infidelity, sexually transmitted diseases, unplanned pregnancies, and heartbreak are always possible when a compulsion includes another human being.

Veterans experience a different kind of trauma than individuals with childhood or adolescent trauma. There are differentiations in trauma typologies that are worth describing here. Perhaps the most significant distinction between trauma types is whether it is interpersonal or non-interpersonal in nature. Interpersonal trauma, in which trauma events are personally experienced, has some differing impacts when compared to non-interpersonal trauma events. Non-interpersonal trauma could be a life-threatening car accident, observation of neighborhood violence, or starvation and other significant deprivation due to war or famine. Interpersonal trauma is typically abuse, neglect, experiencing violence, or observation of a loved one being brutalized or abused. From these two general categories, subcategories of acute, chronic, sustained, periodic, or singular cadence further complicates outcomes, as does age, culture, and gender. For example, a young girl sexually abused by a family member over a period of five years may present with depression, anxiety, and patterns of codependence with partners. A female teen sexually assaulted once may present with more compulsive behaviors, increased startle response, anxiety, and agoraphobic symptoms. The original threat augments the later responses.

For combat Veterans, the trauma experience is far more complex. Recall the definition of *trauma*: inordinate and toxic levels of stress, augmenting the neurobiological processes of the individuals that experience it. Veterans experience trauma in multiple ways--interpersonal violence, observed violence, privation, mortal danger, loss of brothers in service, and the horrors of combat observed as they impact civilians. Observing children blown to pieces by IEDs and women and children routinely raped and discarded are events that are not given time or space to process or even avoid in repetition. Many American soldiers in the Middle East are told routinely to "stand down" when confronted with the commonplace child rape and sexual slavery embodied in public auctions and subjugation of children. Added are the impacts of exacting violence—not simply observing, but following orders and training to take the lives of those that could derail the mission. Combine this with returning to a society that does not understand your experience, a spouse that has a front-row seat to night terrors and behavioral health issues, and civilian relationships that feel empty, disloyal, pointless, and largely unfulfilling. This level of complexity requires an equally complex solution, in the sense that the number of holes in the dam are so scattered and so numerous that combat Vets quickly run out of fingers to plug them.

Complex post-traumatic stress disorder (CPTSD) is often treated with a whack-a-mole approach. Alcoholism may be masking night terrors, leading to poor sleep. That poor sleep is exacerbating high-reactive emotional responses to stressors, and frustration over reactivity leads to marital stress, which leads to infidelity or domestic conflict. Factor in potential criminal justice involvement and loss of employment, and sprinkle in the host of chronic physical issues from untreated injuries and accelerated physical signs of aging. Addressing one issue unmasks another, and many behaviors are simply attempts to self-medicate and avoid dysregulation and trauma-related symptom expression. This is why combat Veterans, particularly multiple deployment Veterans, have such a difficult time truly identifying and treating behavioral health issues. The complexity and intensity in which original threats manifest present equally complex and intense symptoms. With each deployment and re-acclimation home, these stressors compound into increasingly challenging impacts to health and wellbeing. A similar correlative perspective can be observed in violent urban ghetto settings, including privation, observed violence, experienced violence, police brutality, gang violence, poor education, drug abuse, incarcerated family members, unemployment, and poor medical care and access create equally complex traumas. This is where we often see generational, cyclical trauma that seems all but impossible to address given the plethora of uncontrollable issues. Where there is abject poverty, there is violence and trauma.

We are a nation built from behind the muzzle of a gun--regardless of what we believe of ourselves and our histories, we are a country originated in revolutionary violence, divided by civil war, reunited through force, and now act as a post-Eisenhower key player in the global military industrial complex. We are accustomed to the *idea* of war, yet the percentage of combat Veterans to civilians has decreased precipitously in the last 100 years. In World War II, over 11% of the US population was enlisted in the military, with percentages varying based upon the conflict and historic drafts. Currently, less than 0.5% of the US are active duty military--and even less are elite, specialized forces. Yet the celebratory approach to Veterans, the consistent veneration of tongues, if not in practice, rarely discusses honoring Vets by withdrawing from proxy, resource, and ideological battlefronts. With so few people in the US directly impacted by the ravages of war in the modern era, the understanding of its impact on the human psyche becomes less and less prioritized, despite the optics of celebration and veneration.

This is never more apparent than the treatment offered and accessible to Veterans when they leave the service. For combat Veterans returning to civilian life, choosing to live in remote and rural areas for their own health and wellbeing comes with the cost of limited or poorly provided medical and behavioral health care. While many VA hospitals in urban and suburban areas have comprehensive healthcare, rural Vets are tasked with driving 1, 2, even 3 hours to these facilities, making consistent care for chronic issues hugely difficult. The dysregulation as a result of complex trauma and PTSD severely limit Veteran ability to maintain healthy relationships, exercise routinely, maintain employment, downregulate stress response, limit drug and alcohol use, and maintain and organize multiple health concerns. Asking Veterans to adhere to treatment plans that are either insufficient or overly complicated is asking for noncompliance, in the sense that the tools needed to do so are mired in the disrupted neurobiology of stress response that they are inaccessible. Post-service care gets even more complex with degenerative conditions, compounded injuries, age-related physical deterioration, and management of benefits, disability ratings, and poor continuums of care.

The complexity of trauma, dysregulation, and management of chronic, often treatment-resistant diseases and disorders are of course well-met with psychedelic-assisted therapy. Issues of substance abuse and addiction, even for Veterans able to get fully sober and abstain, are markers of underlying traumas often left unresolved with sobriety. Persistent painful memory recall, flashbacks, night terrors, unexplained angry outbursts, compulsive sexual behavior, major depression, and other symptoms of CPTSD are all dysregulatory features of hijacked stress response. The neurogenesis offered by PAT to reset those neural networks through improved and neurogenerative cellular health is the complex solution to the complex issue.

If the original threat informs the outcome, from an epidemiological perspective, reverse engineering symptom sequelae will point to the areas in which intervention is most effectively applied. In this case, combat-related trauma disrupts the same neurobiological processes as developmental trauma--but in a much more condensed window with greater severity and complexity. If the complex stress events are traced to neural degeneration, neurogenesis stands out among treatments as the optimal place to start the healing process.

The Federal illegality of psychedelics is asinine--but stupid quickly becomes harmful when we see psychedelics as highly effective treatments being withheld from combat Veterans we have celebrated, trained, armed, and forced into traumatic environments. The US government sends our men to specialized operations in combat zones, effectively ensuring traumatic experience; it then criminalizes the most effective treatment for managing the impacts of that trauma. It is at best heartbreakingly shortsighted, and at worst, it is abusive in the most insidious and horrific way. We make 'em and break 'em, and then refuse them effective care and threaten their benefits for pursuing such care. It stinks in the way ultra-processed food companies promote obesity, and then sell the medical equipment needed to monitor blood sugar from obesity-derived diabetes. We are societally creating injury and offering masking agents as treatment. It's unacceptable. It's anti-American.

It's worth sharing a story about one man, a multi-deployment special operations Veteran, who was fortunate enough to have the ability to go to South America for ayahuasca treatment. An interview as part of a study on combat trauma was one of the original case studies I'd read with intrigue early in my research. In the year following his honorable discharge, his marriage was in shambles, his alcoholism was increasingly problematic, he'd been cited for DUI, he was battling suicidal thoughts on a routine basis, was unable to maintain consistent employment, and found his friendships and family relationships that once fulfilled him felt hollow and exhausting, including his relationship with his children. He had tried traditional therapy and counseling, medications ranging from SSRIs to benzodiazepines. Each time he would see a therapist, he'd have a new diagnosis--generalized anxiety disorder, major depression, PTSD, and unspecified sleep disorders. His chronic neck and shoulder issues from combat-related injuries, in addition to tinnitus and hearing loss, created ongoing pain and discomfort. The VA doctor he saw at a disability rating appointment was more than willing to prescribe, and did so with consistency and impunity, even after he was suddenly cut off from OxyContin as a result of tightened regulations as opioid addiction ran rampant across the US. He was 28, with chronic pain, a broken marriage, unemployed, suicidal, and had been using heroin for two months when his wife gave him an ultimatum. He does not come back in the house, where he's obviously in active addiction in front of their kids, until he gets his shit together. "It was the worst thing in the world, and the best thing in the world. She loved me enough to give me that chance, and she knew I needed a hard line. She changed the locks, even. It was horrible, and necessary." He remembered considering selling his wife's wedding ring when going through the nastiest part of withdrawal, and getting a text from a buddy concurrently with that thought.

His friend and fellow Veteran had been on many of the same missions, though with a shorter contract. He had struggled for years as well, losing custody of his children after his wife left with them to a different state, which he couldn't even bring himself to fight given his deteriorated state. He had been invited to a retreat in Costa Rica, where he stayed for nine days. He had done an ayahuasca ceremony, engaged in pre- and post-work with the shaman at the facility, and returned home utterly changed. His friend had been on his mind, and miraculously, sent him the information on the program at the lowest point in his life. He equates it to divine intervention.

Somehow, he pushed through the withdrawal, and never did pawn his wife's ring. He called his friend, and was overwhelmed with the release obtained by simply being honest with someone who understood his situation a little. He talked to his wife. "I didn't think she'd be okay with it. I thought for sure she'd think I was lying, or crazy, based on the last year or so. But I think she knew how desperate I was. She was all in."

He was gone for ten days. He self-reported intense, cathartic experiences; vomiting, profuse sweating, and an ability to see behind the fabric of life. He reported being overwhelmed with an entity or being full of love, light, and creativity. He struggled to describe the experience following it, though the impact was significant and life-changing. He returned home with a deep sense of reprioritization, renewed love and appreciation for his wife and children, a deep sense of compassion for not only himself, but all the relationships and lives around him. His wife, initially hesitant to believe in such a significant change's ability to sustain itself, is now a strong advocate for PAT legalization in the US. She describes the difference from departure to return as a light switch--noting that he was completely clean and sober still by the time of their interviews, three years after his ayahuasca journey. He has started his own small business, manages his chronic pain without prescription medication, has repaired and improved his marriage, is an active, devoted father, and no longer meets diagnostic criteria for *any* of his previous diagnoses. He attributes this, entirely, to his ayahuasca treatment.

Not all Veterans will be miraculously cured with a singular psychedelic experience--and this is absolutely not the purpose of this chapter to tout a silver bullet treatment for all Vets. It is ludicrous, however, to withhold the opportunity for miraculous healing when it is *possible*.

Across the board, Veterans with complex PTSD display the same symptomatology as individuals with severe developmental trauma, with the added hardship of being trained to never ask for help. To rub some dirt on it, and go on your way. Pick up your damn ruck. Sleep when you're dead. Eat some other day. Find the target, execute the task. There is no time to be a broke-dick, with severe consequences for being labeled one.

The stories about Veterans finding a new life following PAT are increasing in number, yet we remain stuck in the intractable space of cost-prohibitive, improperly understood provision of psychedelic-assisted therapy. If we, a nation borne from the sacrifice of soldiers, cannot find a way to facilitate this treatment effectively and equitably for our Veterans, we are dishonoring their sacrifice. If we value them only when they are whole and appear invincible, we are not valuing them as people. We are celebrating heroism and neglecting the hero. This is why our next and final chapter will read very much like a psychedelic manifesto: a call to action to make sure PAT has adequate opportunity to do its beautiful work--legally, effectively, and compassionately.

11

A CALL TO ACTION

By this point, you understand that psychedelics--particularly LSD, psilocybin, peyote, and ayahuasca--are not addictive, almost impossible to overdose on, and have significant value in the realm of healing. These three aspects must, absolutely must, move these medicines out of criminalization and inappropriate Schedule I designation. Legalization of psychedelics opens the door to individually-organized use, personal treatment planning, and supervised therapeutic settings. As I write this in the tail end of 2024, I see 2025 and the new administration potentially moving the needle toward widespread legalization. I am hopeful, and I am also cautious: regulation of healing remedies can create structure, and structures can be barriers. For this treatment type, honoring the process becomes of utmost importance.

The deeply individualized experiences during PAT are hyper-specific to the person within the phenomena. This is an issue with the concept of trials, as evidenced in the recent MDMA clinical study in which the study used the "gold standard" double-blind methodology to limited and disastrous results. For patients beginning the process of PAT, work on the front and back end of the sessions themselves is as crucial to success as the medicine facilitating the internal work. This was emphasized repeatedly in the research I conducted--every participant reiterated that the pre-work of training yourself to be aware and open, of intention-setting without bringing strong expectations, and having deep trust in their clinician were monumentally important. This helped during particularly challenging or frightening moments in their sessions, as a gentle touch on the top of their feet would reground them, and belief that their clinician truly was there to keep them safe allowed them to do hard psychological work. Following the sessions, they would journal, debrief, relax, reflect, and move slowly from their dissociated state back to reality. Contrast this with the double-blind scenario. Consider first the mindset of people that are willing to engage in trials of this nature, and the utter lack of pre-work and intention setting, in addition to the clinical environment. Worse still, there was no opportunity for trust development or safety assurance with a trusted clinician--in fact, not even knowing if you took the MDMA begins the trial itself on dishonest terms. It is little wonder people experienced negative outcomes from that trial and those like it. It's frankly *surprising* positive results occurred in any way, for anyone, at all.

Oregon and Colorado have legalized therapeutic use of psilocybin and a few other psychedelics, and Missouri has legalized psilocybin for Veterans suffering from PTSD. These three states also have municipalities that may or may not permit psilocybin use, and federally, it remains Schedule I. This disjointed setting for PAT is a starting place, and it is critical that we use this infancy stage as one in which we move steadily and slowly in the right direction. Slow is smooth, and smooth is fast--yet concerns over the general lack of knowledge around PAT is a significant risk factor.

The heavy weight of terrible information continues to plague public perception of psychedelics, and as with all things, ignorance exacerbates such perceptions. There are a few sticking points for people when it comes to PAT, rooted primarily in the belief that they will unmask or create lifelong psychosis, that people will place themselves in fatal scenarios as a result of being "tripped out", that a "bad trip" will be permanent, that they are addictive, that they will lead to further drug use, and that they will be recreationally abused by young people with lasting consequences. Part of this is the naming issue--psychedelics as a term evokes a ton of 1960's and '70's imagery and associated Manson murders, psychosis, and orgy-like hedonistic concepts. Using terms like *magic mushrooms* and *acid*, despite holding reality-based information (they are indeed mushrooms and lysergic acid diethylamide), paints pictures that are far more uncomfortable than an orange prescription bottle with a prescriber's name and paperwork with a follow-up therapist appointment.

Here is one of your first calls to action, should you choose to do so. Be bold, and spread *factual information* in the same way you would any socially-accepted treatment. One approach I use is to appropriately label psychedelics as neurotechnologies and agents of change; similarly, I refer to them as neurogenerative tools promoting cellular health and resilience. I talk about how they promote healthy synaptic connections, and how they enhance brain plasticity, particularly for individuals with synaptic pruning and degeneration due to trauma, stress, and illness. I also make sure to add that the psychedelics themselves are not what heals, but their ability to open the mind to help individuals work through their experiences in meaningful ways that promote healing from the inside out. Communicate, whenever possible, that overdoses on LSD, psilocybin, and ayahuasca are so incredibly rare that there is greater risk of overdose on allergy medication than on psychedelics. If it doesn't make you utterly intolerable, take a minute to ask people if they know the qualifiers for a drug to be Schedule I, and point to the disconnect between those qualifiers and psychedelics.

It is imperative that, as we ideally move toward widespread legalization, therapeutic settings do not become inaccessible to providers or patients. The requirements for healing centers in Colorado, for example, are similar to requirements for weed dispensaries--video monitoring, limited supplies, rigorous testing, costly permitting, and subject to municipal regulations. This is cost-prohibitive from a provider perspective, and added to that is the minefield of insurance providers, of whom only a few, if any, will consider partially covering or reimbursing for ketamine-based PAT. Despite being legalized for use federally, it is deemed "experimental". Without insurance coverage, providers and facilitators of PAT will be looking at private pay only for revenue, making it increasingly difficult to provide access to individuals that are not already financially solvent with expendable income. Psychedelic-assisted therapy hovers around $6,000 for six sessions of ketamine, with huge variation on either side of that price point. Struggling people rarely have an extra six grand laying around.

The second action item here is to support legalization to force insurance providers to cover this therapy. Without legalization, there is no impetus to cover it, and without a need to do so, healing centers and opportunities to provide therapy will struggle to remain financially viable. This will place individuals with trauma-derived chronic disorders in the two camps that already harm Americans in our existing healthcare system: those with money have access to better health. Those without suffer the consequences. The Veterans Administration stands to win the most by widespread legalization and benefits covering this treatment. Chronic, treatment-resistant disease, suicide, and violence are costly to individuals and society, and for the VA, covering a treatment that has high efficacy and low risk pencils out financially far more easily than paying for medical costs associated with addiction, unemployment, mental illness, and physical and cognitive deterioration. Over time, healthy people cost far less to insure and care for. Yet here is the next barrier in widespread use of PAT: our existing definition of *healthcare* must fundamentally change.

Currently, healthcare in the United States is based upon employer-sponsored health insurance plans and state-subsidized insurance. It is all, regardless of category, expensive and insufficient to cover all needs. Insurance providers dictate care, which is a point of disgust in a society ostensibly meant to support human thriving. A cancer diagnosis, even with insurance, can destroy a family financially within months. Healthcare in the US is not about caring for your health, it's about pharmaceutical and insurance corporations extracting every cent they can from human beings with impunity. Psychedelic-assisted therapy poses a huge threat to these industries, as a person no longer trapped in chronic diagnosis management spends far less on prescriptions, treatments, and insurance premiums. In the same way nutritional health diminishes profits for insurers and drug companies, psychological health will impact bottom lines as well. It is imperative that PAT and the natural medicines within it remain outside the greedy machinations of corporate interests. The real maleficence is in the insidious way these industries promote their "cures" and "healthcare provision". Consider the widespread use of Ozempic and its sister drugs. The issue is obesity--underlying causes of metabolic dysfunction, compulsive eating, and ultra-processed food impacts provide a visual representation through overweight. Rather than addressing metabolic disorders, poor food quality, and overconsumption, Ozempic acts upon the satiation mechanism, informing the brain that the body is not hungry. The results are dramatic weight loss, diabetes reversal, and diminished physical issues as a result of healthier BMI. These can be achieved *without* the drug--but the cure is now ripe for dependence, because as soon as you stop taking the drug, the satiety signal reverts back to its old self in the absence of the mimicked GLP-1 hormone hunger suppressor. Rather than using these glucagon-like peptide-1 receptor agonists as *beginning* treatments, complete with titration *off of them* as the treatment plan, the intention of the industry is lifelong purchase by not addressing the root cause. This is a decidedly important consideration in the widespread legalization of psychedelics--corporate interests co-opting "cures" leads to dependence on those cures. I offer caution now regarding large scale manufacturing of psychedelics for prescription use, should they become federally legalized: the intention behind PAT is to be free from the trappings of chronic disease and disorder. Placing all production and testing in the hands of disreputable and dishonest drug companies, and restricting access to treatment in the hands of integrity-challenged insurers will prevent individually obtained effective care. A lengthier call to action: refute any and all corporate gatekeeping in its many guises. Exorbitant costs, selective network providers, and denial of claims are standard extortion practices by drug and insurance corporations--keep PAT far, far from these entities.

Finally, I ask that you support the widespread provision of PAT for Veterans however *they* prefer it. Whether it's a facilitated experience in their living room, a stay at an inpatient clinic, or a guided journey in the woods on a retreat, support it. Insist upon it as a part of benefits provided by the VA, and support the expansion of services in rural areas to allow for easier access to care. Importantly, support reimbursement by the VA to any and all providers licensed to facilitate PAT. There must be no barrier--financial or otherwise--to providing this treatment to Veterans. It is not a treatment readily or reasonably abused; it is non-addictive; it facilitates increased efficacy of other treatments, and provides the absolutely crucial neurobiological support to give Veterans the tools necessary to build back a life of health and meaning. The decision-making on PAT should be in the hands of those who understand psychedelics, recovery, and regulatory health. It should also be in the hands of those who benefit from it the most--Veterans, and anyone struggling through life desperate for change and unable to grasp it through the fog of trauma-induced dysregulation.

Here's your action list:

1. Inform others with facts, kindly, about psychedelics and their role in recovery. Give them this book. If cost prevents it, email the address at the end of this book.
2. Support federal legalization and decriminalization of psychedelics. This will lay the groundwork for requiring insurers to cover treatment.
3. Refute corporate strangleholds on psychedelic production and utilization through supporting responsible home cultivation and community education.
4. Write to your representatives encouraging them to advocate for VA benefits to cover PAT in perpetuity for Veterans, without barriers to access based upon referrals and diagnoses. We trust our Vets to kill on our behalf--trust them to know what treatment will work for them.

Psychedelics have been used throughout human history as conduits of faith, culture, redemption, healing, communion, maturation rites, and celebrations. The explosive nature of human development is largely attributable to nutrition, and if one looks at the impact of food on the human brain, we must also look at the impact of psychedelics on the human mind. The ability to recover neurological injury and cellular death, the critical consideration of new ideas and concepts otherwise unobservable, the incredible reconnection to the world and all those within it...all of these facilitate human thriving. The acceptance of death and dying through PAT, which has been studied with increasing breadth in the last fifteen years, speaks to the unique opening psychedelics provide to minds trapped in oppressive, cyclical thoughts. People with terminal diagnoses, whether during the 1990's HIV/AIDS epidemic or a late-stage cancer diagnosis, are faced with the terrible reality of death. The fear of being pulled into darkness is a permeating reality we see woven throughout human history. Religion is our management of knowing we will die, and our adorable attempt to create binding properties around death that we often believe with every fiber of our being. Yet when the moment comes, if we see it coming, we are still overwhelmed with the unknown, regardless of our strength of faith. While we didn't cover it much in this book, the experiences of terminally ill patients during and following PAT summarize what I hope you now believe, too. What's next is beautiful, and what's now is beautiful, and what you are is worthy of love.

In patients given a terminal diagnosis, a population deeply immersed in navigation of meaning (particularly for those younger and middle aged), the diagnosis of *early death* was deeply traumatic. The loss of life meant the loss of family, leaving loved ones, and watching with increasing anxiety the deterioration of their physical bodies. Deep depression over things left undone, future missed moments, and the stark reality that no one knows what happens when the light goes off compound into a massive, malignant state of despair. Unlike other issues, a terminal diagnosis does not provide *any* notes about recovery and future health.

It's all about how bad it's going to be before it all ends, in a final and unwanted chapter of death and dying. Suicidal ideation is incredibly common, as is overuse of pain medication, drugs and alcohol, and fracturing of family ties due to stress and hardship unavoidable in such diagnoses. It is a dark and troubling way to exit your life. A few more notes are important to expand upon regarding end-of-life struggles.

For patients reporting all of these fatal trauma-related fears, hardships, and desperate states of being, PAT may seem odd as a treatment option. It won't treat the illness--that is stuck. The patients who undergo PAT in these contexts reveal the same underlying themes elicited in those recovering from trauma-related chronic disorders. Patients reported, from PAT until death, a reversal in depression and anxiety. They reported elimination of anger and resentment at the injustice of early death, and drastically improved relationships with loved ones despite deteriorating health and the stress that accompanies it. Patients confirmed, repeatedly, that they knew this life was not the end, and that death was a door to the next state of being. They report peaceful acceptance of their own mortality, and joy in the belief that each person in the world has value in their lives, no matter the years provided to them, and experienced a tremendous connection to each person alive today--simply through shared humanity and radical compassion. Terminally ill patients once frantic with the idea of inescapable death reported immediate relief through the ability to see their illness as one small feature of a massive life-force undulating with love and regeneration. This is the key, and it's what each person at the end of all options reaffirms as a result of psychedelic-assisted therapy. It's what is missing in America from our fractured identities, unhealthy physical states, and ineffectively treated individual and societal traumas. It's the sense, the knowing, that you are important. You are *enough*. You *are* worthy of love.

It is beyond inconceivable to me that this medicine is withheld from Americans, touted as dangerous, made a vehicle for incarcerating those in possession of it, and demonized for actions and ideologies that were never a part of its history. Psychedelic-assisted therapy is a means of understanding the self, and the self in relation to lived experiences and the worlds you occupy. It is particularly fascinating as a conduit for time travel--as strange as it seems, this is one of the most healing attributes of psychedelics. Memory is a powerful tool for health *and* harm. Like our anything-is-poison-if-overdone concept we talked about earlier, being unable to escape memories that alter our present state of mind is an enormous hardship for people with painful memories. When your mind is no longer yours to command, and time is no longer consistent or predictable, you live with emotions that arrive without warning and without regard for your needs or wants. The dissociative aspects of psychedelics permit examination of memories without being attached and influenced by them. In conflict intervention and mediation, this is often called *the balcony perspective*. This term is usually used to resolve conflicts between parties by taking the person out of the issue and looking down on the issue together, from your respective balconies. You can put binding aspects on the conflict--what it truly is, why you were responding as you did, where the similarities lie in your perspectives, and what approach might be best to tackle THE problem, not YOUR problem. Psychedelics permit this within individuals, allowing them to suspend themselves from time and emotions to examine the issues interrupting their happiness with a critical consideration. This is why those who have experienced the dramatic recovery possible with PAT describe seeing issues in a different way, being surprised at who and what they encounter, pulling solutions and new ideas out of textures and patterns, encountering themselves as children and bringing that encounter through time to the present, and more. This is why death is less frightening, too--there is a sense that you are no longer trapped and bound by your existence, but rather, you are a part of the fabric of everything. Being overcome with firm knowledge that the universe is an undulating, creative, energetic force helps knit together the tears in its connective fabric to your unique human experience. The most common words and phrases from some of the most broken and hurting people post-therapy are *hope, love, creativity, strength, connection, beauty,* and *safety.* This is healing--this is epiphany, this is recovery, this is *magic.* And it must be made available to those who need it.

Imagine with me, in these last moments we are sharing in this literary conversation, a world with widespread understanding and acceptance of psychedelics as a healing mechanism. In the United States, we see rampant, unchecked brokenness in our governmental, educational, medical, social, and religious activities. Our government sends our citizens to kill and be killed with the same blithe pen stroke in which a celebrity provides an autograph to a fan. Our public schools are overrun with deficient reading skills, inattentive children, and curriculum standards set by textbook publishers that have no interest in active, dynamic, experiential education. This makes our teachers and students the unfortunate victims of industrial-era factory grading requirements, diminishing interest, broken enthusiasm, and limited love of learning within the schools and therefore society. Universities become institutions of political ideologies, often connected to grant funding and enrollment needs, augmenting hearts and minds to the latest ideology instead of the practice of critical thought. Our medical care is in the hands of corporations, setting standards of care based upon financial benefit to shareholders, without regard for any individual's poverty, addiction, illness, and death. Socially, we are bred to consume--every holiday, every event, every idea, every change, *everything* must be commodified to enable extraction into the market. This extends from identity politics around family life to the very abuse and trafficking of children secretly and subversively for the pleasure of the highest bidder. Our religious health is fractured, too--our religions become pithy sayings on our living room walls, temporary banners on social media profiles, and reasons to disagree at Thanksgiving, rather than guiding principles of community, accountability, and morality. What if, though, *what if*, we were able to disconnect from our painful pursuit of filling endless holes? What if we could, just for a moment, stop the hubbub and high-pitched whine of industry and avoidance to access a moment of reflection?

This is an opportunity to reprioritize our values as Americans. This is the moment we can push pause for just a *second* on limited solutions and truly reflect on the human experience we are navigating. Our brains are working so hard--so very, very hard--doing what they are designed to do. They are designed to take in all of our experiences and environments through our senses, to process stimuli through the operational system of our neurobiology, and provide an output that in turn will augment our realities and trajectories in life. We have so much to navigate as humans, and for so many of us, our civil society has ceased to be a place we feel truly connected. Our families are forced to navigate structures of industry in which people are both the product *and* consumer, making it easier every day to create harm with the dehumanization required to maintain this extractive business. It is essential that each individual within this wonderful country has the opportunity to develop a sense of purpose *from within,* rather than hunting it down through endless scrolling and obsessive sociopolitical hobbyism. Individual traumas, social traumas, even national traumas are all rooted in the removal of self-efficacy, identity, and aspirations, creating endless opportunities for extortion and sadness.

It is my firm desire, my obvious wish, and my authentic plea: psychedelic-assisted therapy must be made available, accessible, legal, and understood. We have the incredible hearts of American dreamers, and within us the ability to truly become exceptional, innovative, independent nation-carriers. We must respond appropriately to the experiences in our society that diminish our ability to grow beyond our diagnoses, our hurts, and our disrupted and dysregulated development. That appropriate response is a deep examination of pathways maladapted to be able to survive in the world, in an effort to reset them to pathways *adapted* to thrive in it.

Giving people the tools and opportunities to heal from within is a calling that is unignorably of highest import. Answering this call means reimagining what health looks like, and refuting claims that we are incapable of full recovery while PAT remains outside common behavioral health solutions.

We are not defined by our trauma.

We are defined by our healing.

So let's heal.

First, thank you. Thank you for reading this and being interested in things that may be uncomfortable or difficult. If this book meant something to you, or to someone you know, or if you'd just like to talk about it, please do not hesitate to reach out. I cannot promise to respond to every email, and like many writers, I occupy a "regular life" of full time employment and family obligations. I *can* promise that I will do my best to respond. If you have questions, comments, feedback, requests, etc., please write to me. If you know someone who could use this book and can't afford it, that's a good reason to write, too.

talkabouttheresetbook@gmail.com

www.ingramcontent.com/pod-product-compliance
Lightning Source LLC
Chambersburg PA
CBHW071713130725
29531CB00036B/1661
9798218590611